Corona Discharge Micromachining for the Synthesis of Nanoparticles

Characterization and Applications

T0141326

Corona Discharge Micromachining for the Synthesis of Nanoparticles

Characterization and Applications

Dr. Ranjeet Kumar Sahu and
Dr. Somashekhar S. Hiremath

CRC Press
Taylor & Francis Group
Boca Raton London New York

CRC Press is an imprint of the
Taylor & Francis Group, an **informa** business

CRC Press
Taylor & Francis Group
6000 Broken Sound Parkway NW, Suite 300
Boca Raton, FL 33487-2742

First issued in paperback 2021

© 2020 by Taylor & Francis Group, LLC
CRC Press is an imprint of Taylor & Francis Group, an Informa business

No claim to original U.S. Government works

ISBN-13: 978-0-367-78798-1 (pbk)
ISBN-13: 978-0-367-22473-8 (hbk)

Visit the Taylor & Francis Web site at
http://www.taylorandfrancis.com

and the CRC Press Web site at
http://www.crcpress.com

Contents

Preface

OVERVIEW

Nanoparticles are cornerstones of nanotechnology which play a significant role this century, due to their enhanced size-dependent characteristics compared to extremely fine or larger particles of the same material. Generally the size of the nanoparticles is less than 100 nm. Their uniqueness arises from their large surface area to volume ratio. Over the last few decades, researchers have gained a significant interest in metallic nanoparticles because of their novel physical, thermal, catalytic, electrical, antibacterial, and optical properties, and also due to a variety of multifarious and interdisciplinary emerging applications. They are used mainly in thermal management systems like heat exchangers and evaporators, MEMS devices, electronic devices, catalysis, rocket propellants, explosives, biomedical drug delivery, food protection, etc. In recent years, there has been significant attraction towards colloidal suspensions (nanofluids). Nanofluids consist of solid nanoparticles suspended in a base fluid and are used for various applications. In most of the studies the common base fluids include water and organic liquids.

To date a wide variety of synthesis methods have been developed for synthesis of nanoparticles of required sizes and shapes. These methods include ball milling, liquid reduction, sol-gel, micro-emulsion, wire explosion, sputtering, laser ablation, combustion flame, etc. But, the generation of nanoparticles using corona discharge micromachining—Electrical Discharge Micromachining (EDMM)—has not been explored and adopted in the present investigation.

The EDMM method has been used for machining micro-features like micro holes and micro channels on various engineering materials. Some of the notable applications are micro holes in nozzles, filters, turbine blades, catheters, needles, and micro channels in reactors, MEMS, and fluidic devices, etc. These micro-features are obtained by removing a tiny bit of material in the form of debris/chips using repetitive spark discharges which appear between the tool (cathode) and the workpiece (anode), immersed in a dielectric medium. It is quite interesting to note that the debris/chips generated during the EDMM process are very small—micron/nano size—and the exploration of this debris and its applications as nanoparticles in the field of nanotechnology have yet to be addressed. It is possible to control the size of this debris using controlled process parameters. By closely controlling the EDMM process parameters namely voltage, current, frequency, duty cycle, and pulse duration, it is possible to generate a large quantity of nanoparticles of required size and morphology. The main features of an indigenously developed EDMM prototype system consist of workpiece electrode—copper and aluminium plate; tool electrode—copper and aluminium wires; dielectric fluid—De-Ionized (DI) water; tool feed—piezoactuator; control circuits—pulse generation circuit; and tool feed control circuit and ultrasonicator.

Experiments were conducted using an indigenously developed EDMM prototype system to synthesize copper and aluminium colloidal nanoparticles. The characterizations of colloidal suspensions using various diagnostic studies such as Transmission Electron Microscope (TEM), Energy Dispersive Analysis by X-Rays (EDAX), and Selected Area Electron Diffraction (SAED) pattern were carried out to ascertain the size, shape, and elemental composition of the colloidal suspensions.

Initial experiments conducted using only DI water reveals that large sized particles are seen during TEM analysis and found that this is due to the agglomeration of the particles. To overcome this problem, the various stabilizers are used along with DI water as Poly-Vinyl Alcohol (PVA) and Poly-Ethylene Glycol (PEG) for copper nanoparticles and Poly-Ethylene Glycol (PEG), Bael Gum (BG), and ACacia Gum (ACG) for aluminium nanoparticles.

The stabilizer PEG when added to the DI water was found to have a major effect on the reduction in the mean size of copper particles as compared to the PVA polymer. Similarly, the mean size of the aluminium particles was estimated to be less for those generated in ACG solution when compared to that in DI water with PEG and BG polymers. A novel ultrasonic velocity technique has been used for the concentration characterization of the synthesized colloidal suspensions.

Thermal conductivity analysis has been carried out using a KD-2 Pro device. It was observed that the thermal conductivity of generated colloidal suspensions has enhanced with respect to base fluid DI water. Viscosity analysis has been carried out using a Brookfield viscometer. The viscosity of the colloidal suspensions was estimated to be higher than that of pure DI water.

Ultraviolet-visible (UV-vis) spectra analysis has been carried out using a UV-vis spectrophotometer. This analysis was used to ascertain the nature of the nanoparticles in the solution after keeping the samples for longer duration. Formation of oxide layers on the generated particles was observed. Finally, a thermal management experimental setup has been developed to test the application suitability of generated colloidal nanoparticles for heat transfer. The experimentation reveals that the colloidal suspension of copper has the ability to transfer more heat in the existing heat transfer system when compared to colloidal suspension of aluminium.

STRUCTURE OF THE BOOK

The aim of this book is to summarize the fundamentals and established methods of synthesis and characterization of nanoparticles, so as to provide readers with an organized and rational representation of the field. The readers will readily find that this book has focused primarily on a recently explored corona discharge micromachining method to synthesize inorganic nanoparticles and of course, in the synthesis of nanoparticles, organic materials often play an indispensable role, such as polymers in the form of polymer stabilizers. The book will also be devoted to characterization and applications of nanoparticles synthesized via EDMM. Therefore, this book would be well suited to upper-level undergraduate and graduate courses as well as serve as a general introduction to people just entering the field, and would be sufficiently readable and thorough so that it can reach academicians who have a need to know the nature and prospects for the field of nanoparticle technology. However, as nanoparticles research is evolving and expanding very rapidly it becomes difficult to cover all the aspects of the nanoparticle technology field in this book.

Synthesis of nanoparticles is the first key step to investigate the application characteristics of nanoparticles. Therefore, Chapter 2 has been devoted to a discussion on various established methods used for synthesis of nanoparticles. This will provide readers with a systematic and coherent picture about synthesis methods of nanoparticles.

In order to realize the properties of nanoparticles, there is a requirement to study various diagnostic methods, which can be an important aspect of this book. Chapter 3 is focused on the description of various diagnostic methods used for nanoparticles characterization. The methods include Scanning Electron Microscopy (SEM), Transmission Electron Microscopy (TEM), Energy Dispersive Analysis by X-Rays (EDAX), Selected Area Electron Diffraction (SAED), X-Ray Diffraction (XRD), scanning probe microscopy, ultraviolet-visible spectroscopy, ultrasound velocity method, thermal and electrical conductivity measurement, and viscosity measurement methods.

In Chapter 4 the need for EDMM and the EDMM approach for synthesis of nanoparticles are discussed first. The development and fabrication of EDMM experimental setup coupled with a sonicator and piezoactuated tool feed system are then discussed.

The significance of copper and aluminium nanoparticles in the current situation is discussed in Chapter 5. The methodology of carrying out the experiments for synthesis of copper and aluminium nanoparticles using the developed prototype EDMM system is explored. The characterization results of the synthesized particles and the possible methods to overcome agglomeration of particles and achieve stable dispersion are discussed. Further, the application suitability of synthesized nanoparticles is examined. This includes the development of a thermal management system for heat transfer application of colloidal particles. Conclusions and future directions are listed so that the intended audience can obtain an appreciation of developments outside their current areas of discipline and can obtain an overview of the book.

Notations

ENGLISH SYMBOLS

v	Velocity of sound
k	Adiabatic compressibility
v_s	Ultrasonic velocity in the suspensions
w	Mass fraction
K	Thermal conductivity
q	Electrical power applied
L	Needle length
T	Temperature of the needle
t	Time
n	Order of scattering
N	Spindle speed
T_k	Spring torque constant
T_v	Viscometer torque (%)
d	Lattice spacing
T	Acoustic time of flight measured at different positions of probe
T_s	Average of the measured acoustic time of flight between the successive positions of probe
h	Length of the fluid column
V	Sedimentation velocity
D	Diameter of the particle
M_o	Mass of the incident ion
M	Mass of the target atom
E_o	Energy of the incident ion

GREEK SYMBOLS

μ	Micrometer
β	Bulk modulus
ρ	Density
φ	Volume fraction
$\dot{\gamma}$	Shear rate
τ	Shear stress
μ_d	Dynamic viscosity
λ	Wavelength
θ	Diffractions angle

CHAPTER 1

Introduction

1.1 FUNDAMENTALS OF NANOPARTICLES

In this millennium, nanoparticle technology has become one of the most important and exciting forefront fields in physics, chemistry, engineering, biology, etc. It shows great scientific promise as it provides many breakthroughs at present and will change the direction of technological advances in a wide range of applications in the future. Nanoparticles are microscopic particles at the size of typically less than 100 nm. They are efficiently linked between molecules and bulk materials in such a manner wherein molecules (size range: 0.1 nm to 1 nm) contain $1 \leq$ atoms $\leq 10^2$, nanoparticles (size range: 0.1 nm to 100 nm) contain $10^2 \leq$ atoms $\leq 10^6$, and bulk materials (size range: >100 nm) contain $>10^6$ atoms (Poole and Owens, 2003).

This century, nanoparticles play a considerable role in the scientific, industrial, and medical fields due to their increased size-dependent characteristics compared to larger particles of the same material. Their uniqueness arises from their high free surface energy indicating the presence of a large fraction of atoms on the particles surface that are chemically unsaturated (Gleiter, 2000; Hosokawa et al., 2012). Table 1.1 shows the correlation between particular sized nanoparticles and the percentage of

TABLE 1.1   Correlation between Particular Sized
Nanoparticles and the Percentage of Atoms Present
on the Particle Surface over its Whole Volume

Size of particles (nm)	Number of atoms on the surface (a)	Number of atoms over the whole volume (b)	% (a) with respect to (b)
2	488	1000	48.8
20	58,800	1×10^6	5.9
200	6×10^6	1×10^9	0.6
2×10^3	6×10^8	1×10^{12}	0.06
2×10^4	6×10^{10}	1×10^{15}	0.006

atoms present on the particle surface over its whole volume. It is shown that the fraction of surface atoms for a 2×10^4 nm sized particle over its whole volume is 0.006%, but it increases to 0.6% for a 200 nm sized particle and for a 2 nm sized particle almost 50% of the atoms are found to be present on the surface. From the table, it can be seen that the nanoparticle of size 2 nm will exhibit remarkable characteristics compared to the larger sized particles.

1.2 CLASSIFICATION OF NANOPARTICLES

The spectrum of nanoparticles ranges from inorganic to organic, crystalline to amorphous, aggregates, powders or dispersion in a matrix, suspensions and emulsions, nano-layers and films, class of fullerenes and their derivatives. There are various approaches for classification of nanoparticles. Table 1.2 shows the classification of nanoparticles based on the dimension, phase structure, production process, and nature of atomic bonding (Luther, 2004; Johnston and Wilcoxon, 2012).

The semiconductor nanoparticles are made of those elements (such as silicon, boron, germanium, and carbon) which are semiconductors in the solid state. The bonding between the atoms in such nanoparticles is covalent and the bonds are strong and directional. The molecular nanoparticles can be formed by supersonic expansion of molecular vapor and the type of bonding usually

TABLE 1.2   Classification of Nanoparticles

Classification	Examples
Dimension	
• 3 dimensions <100 nm	• Quantum dots, spheres, etc.
• 2 dimensions <100 nm	• Tubes, fibers, wires, platelets, etc.
• 1 dimensions <100 nm	• Films, coatings, multilayer, etc.
Phase Structure	
• Single-phase solids	• Crystalline, amorphous particles and
• Multi-phase solids	layers, etc.
• Multi-phase systems	• Matrix composites, coated particles, etc.
	• Colloids, aerogels, ferrofluids, etc.
Production Process	
• Mechanical procedure	• Ball milling, lithography, etc.
• Liquid phase reaction	• Sol-gel, precipitation, electrochemical
• Vapor phase reaction	process, etc.
	• Sputtering, laser ablation, flame synthesis, etc.
Nature of Atomic Bonding	
• Covalent bonding	• Semiconductor nanoparticles
• van der Waals bonding	• Molecular nanoparticles
• Ionic bonding	• Ionic nanoparticles
• Metallic bonding	• Metal nanoparticles and nano-alloys
• Coordinate bonding	• Passivated metal nanoparticles

exhibited by molecular nanoparticles is van der Waals bonding. The ionic nanoparticles are electro-statically bonded in nature and are formed between sodium and chlorine in NaCl, magnesium, and oxygen in MgO, and so on. The metallic elements form a wide variety of nanoparticles ranging from the s-block metals (such as the alkali and alkaline earth metals), p-block metals (such as aluminium), and d-block metals (such as transition metals) where the bonding is metallic in nature. Metallic nanoparticles that are composed of more than one metal give rise to nano-alloys. To overcome the handling of the bare metal nanoparticles on a preparative scale and to achieve approximately uniform sized particles, the metallic nanoparticles are protected or passivated with a ligand surfactant shell through the typical coordinate bonding.

1.3 OVERVIEW OF APPLICATION OF NANOPARTICLES

Typically, materials such as metals and non-metals can be modulated into nano-scale dimension in the form of or without colloid. Colloids are nanofluids/colloidal suspensions which consist of nanoparticles suspended in a base fluid (usually water or organic liquids such as oils, glycols, etc.). The colloids can be prepared by combining a large variety of nanoparticles and base fluids. The presence of nanoparticles results in improvement in the properties of colloids that could offer several benefits, which would be significant to various industries. The metallic nanoparticles include copper, aluminium, silver, iron, gold, etc. and have exceptional physical, thermal, electrical, catalytic, optical, and antibacterial properties, etc. Non-metallic nanoparticles used are alumina, copper oxide, titania, iron oxide, zinc oxide, carbon nanotubes, etc. and they have various unique properties such as electron transport, semiconducting property, ferromagnetism, giant magneto resistance, luminescence, ferroelectric property, and catalysis (Hosokawa et al., 2012). The remarkable properties of particles are attributed to small size, surface, and macroscopic quantum tunneling effects in itself, which are unlike from those of bulk metals. Because of which, they call for different applications including electronics, automobiles, catalysis, rocket propellants, explosives, biomedical and drug delivery and food protection, optical fields, sensors, rechargeable batteries, solar cells, pigments, etc. To provide the reader a sense of these applications, a broad overview of the potential applications of metallic and non-metallic nanoparticles is shown in Table 1.3 (Taylor et al., 2013; Saidur et al., 2011; Drelich et al., 2011; Sindhu et al., 2007; Kassaee and Buazar, 2009; Rai et al., 2006; Kulkarni et al., 2008; Basha and Anand, 2011; Hemalatha et al., 2011; Garg et al., 2008; Pfeifer et al., 2005; Johnston and Wilcoxon, 2012).

TABLE 1.3 Broad Overview of the Potential Application of Nanoparticles

Types of Particles	Applications
Metallic Nanoparticles	
Magnetic nanoparticles—iron, nickel, cobalt	Heat transfer, drug delivery, cancer treatment, pollution cleaning, biomedical imaging, flow control, microwave absorbers, sensors, brakes, motors, dampers, seals, bearings, micro fluidic, hyperthermic treatment
Non-magnetic nanoparticles—Cu, Al, Ti, Zn, Pb, etc.	Automobile engines, fuel cells, lubricant oils, protection of food from bacteria, MEMS devices, rocket propellants, explosives, batteries, coatings on glass, solar collectors, electrical contact systems, envelop materials for sodium lamps, inks, heat pipes
Noble nanoparticles—gold, silver, platinum, palladium	Drug delivery, cancer treatment, catalysts, heat transfer enhancement, water purification, photothermal hyperthermia
Semiconductor nanoparticles and quantum dots—Si, Ga, In, ZnS, CdS, CdSe	Fluorescence, light trapping, LEDs, optical memory, solar cells, medical imaging
Core-shell nanoparticles—metals-metals core-shell and metals-semiconductors core-shell	Biomedical, fluorescence, solar absorbers, sensors, optoelectronics, catalysts
Non-Metallic Nanoparticles	
Ceramic nanoparticles—Al_2O_3, TiO_2, ZnO, CuO, Fe_2O_3, WO_3, SiC, SiO_2, AlN, MoS_2, etc.	Automobiles, catalysts, thermal absorption systems, optoelectronics, nuclear reactors, transformer oil, chillers, cameras, micro devices and displays, antibacterial activities, pigments, rechargeable batteries
Carbon based nanoparticles—graphite, graphene, CNTs, diamond	Engine transmission oil, water purification, heat exchangers, domestic refrigerators, electric motor windings, dish towers, sensors for improving oil exploration
Polymer nanoparticles	Biomedical (drug delivery), conductive optoelectronics

1.4 RESEARCH ON NANOPARTICLES

The expansion of nanoparticles in a wide variety of applications has sparked multidisciplinary research and development of nanoparticles. In this section, a detailed review of the synthesis, stabilization, and application characteristics of all types of particles is not covered, but rather the focus is on those nanoparticles which have been gaining immense attention within research and the markets over the last decade and a half.

1.4.1 Review on Nanoparticles Synthesis

The copper nanoparticles in De-Ionized (DI) water synthesized using a submerged arc synthesis method. The mean size of the particles was found to be less than 100 nm. The nanoparticles in DI water were found to be oxidized completely after 1 month (Chang et al., 2000). Thin films of colloids of Ag-Cu alloy and phase separated mixed colloids of Ag and Cu were prepared through a sol-gel route. The alloy colloids were prepared by mixing Cu ions in the presence of hydroxylamine hydrochloride with silver colloids stabilized by polyvinylpyrrolidone, whereas phase separated Ag-Cu mixed colloidal particles are formed by mixing Cu and Ag both as ions. The sizes of Ag-Cu alloy particles are observed in the broad range of 5 nm to 40 nm and the phase separated Ag-Cu mixed colloid are in the range of 8 nm to 20 nm (Suyal, 2003). A chemical reduction method used to produce copper nanofluids by reducing copper sulfate pentahydrate ($CuSO_4.5H_2O$) with sodium hypophosphite monohydrate ($NaH_2PO_2.H_2O$) in Ethylene Glycol (EG) under microwave irradiation. The polyvinylpyrrolidone was used as a stabilizer to retard the growth and agglomeration of nanoparticles, and the particles observed are in the size range of 10 nm to 20 nm with spherical morphology (Zhu et al., 2004b). The nanocrystalline particles of copper are synthesized using the solvated metal atom dispersion technique. The nanoparticles are in the size range of 20 nm to 45 nm. The particles are observed to be more reactive during methanol formation (Ponce and Klabunde, 2005). The mixtures of copper oxide (CuO) and zinc oxide (ZnO) nano-powder with an average size of 41 nm and 77 nm was ball milled to the size of 20 nm. The milled particles are used for catalyst washcoats for micro-structured reactors (Pfeifer et al., 2005). Using the submerged arc synthesis method copper oxide nanoparticles are synthesized in DI water, pure Ethylene Glycol (EG), and DI water and EG mixture at different concentrations of EG. The size and shape of the nanoparticles were found to be influenced by changing the base fluid. This was attributed to the differences in change in the free energy when new phase is formed during nucleation, growth, and condensation, and also due to the characteristics of the base fluids themselves (Lo et al., 2005). The copper nanoparticles of mean size of 140 nm are produced by feeding aqueous solution of copper acetate to the super-critical water reactor. The nanoparticle surface was observed to be highly catalytic during H_2 production by methanol reforming (Gadhe and Gupta, 2007). The copper nanoparticles of mean size 200 nm are produced using the chemical reduction method and then dispersed with EG using a sonicator to formulate colloidal suspensions (Garg et al., 2008). Spherical shaped copper nanoparticles have synthesized using the electrochemical reduction process. The size of the nanoparticles was observed to be in the range of 40 nm to 60 nm and the average size was found to be 45 nm. The spectra analysis of copper nanoparticles shows a plasmonic absorption band at 587 nm, which is attributed to the synthesis of pure particles (Raja et al., 2008). By reducing the power of polyol solutions, copper nanoparticles are produced. The average size of the particles was found to be 21.2 nm, and when polyvinylpyrrolidone was used as a stabilizer the size was reduced to 12.7 nm (Abreo, 2009). The copper and iron nanoparticles were first synthesized individually using chemical precipitation and the average size of the particles was found to be around 500 nm. Then the synthesized particles are dispersed in Ethylene Glycol (EG) via ultrasonic irradiation to produce respective colloidal suspensions (Sinha et al., 2009). The sputtering method was adopted to produce copper nanoparticles of average size of 10 nm. The produced nanoparticles are dispersed in both distilled water and distilled water with 9 wt% of sodium lauryl sulfate stabilizer to study the Critical Heat Flux (CHF) of nanofluids (Kathiravan et al., 2010). The colloidal suspension of copper particles chemically synthesized using boiling flask-3-neck and the microfluidic reactor. In both of the cases, sodium dodecylbenzenesulfonate was used as a stabilizer. The nanoparticles produced through the microfluidic reactor have a mean size of about 3.4 nm with a narrow size distribution, and by the flask method they have a mean size of about 4.9 nm (Zhang et al., 2010). The copper nanoparticles in water-oleic acid mixture were prepared using the two-step chemical reduction route. In the first step, the spherical shaped copper nanoparticles are produced by reducing copper ions from copper sulfate pentahydrate with glucose and sodium hypophosphite in solution. In the second step, copper nanofluid was produced by redispersing nanoparticles in water/oleic acid mixed solvent. The oleic acid is used as a surfactant to prevent nanoparticles from oxidation and agglomeration. The obtained size lies in the range of 20 nm to 40 nm (Wen et al., 2011). Copper and copper oxide nanoparticles were produced by exploding copper wire (current density 10^6 A/cm^2 and voltage 10 kV) in the various ambient air pressures at 1 bar, 500 mbar, 100 mbar, and 50 mbar. When the pressure is reduced from 1 bar to 50 mbar, the size of the particles was found to be reduced from 31.3 nm to 23.6 nm and size distribution of the nanoparticles follows log-normal distribution (Lee et al., 2012). Using the single-step chemical reduction method EG based colloidal suspension of copper particles with an average size of 7.3 nm was synthesized. The particles are synthesized by reducing copper ions from copper nitrate hydrate [$Cu(NO_3)_2.H_2O$] with sodium hypophosphite monohydrate ($NaH_2PO_2.H_2O$) as a reducing agent in EG solution, in the presence of a polyvinylpyrrolidone stabilizer under microwave irradiation (Robertis et al., 2012). The nanosize copper and cadmium chalcogenides synthesized by the mechano-chemical route from elementary precursors using a ball mill. The spherical shaped copper chalcogenides and hexagonal cadmium chalcogenides lie in the range of 8 nm to 19 nm (Kristal et al., 2013). Aluminium nanoparticles are produced using the wire explosion process in inert ambience. The particles size was found to be in the range of 100 nm to 200 nm (Dokhan et al., 2002). The alumina nanoparticles are synthesized by combustion of a small sample of solid rocket propellant. The nanoparticles form chain-like aggregates of size about 1000 nm are composed of small spherical primary particles which lie in the range of 10 nm to 150 nm (Karasev et al., 2004). The wire explosion method was adopted to generate aluminium nanoparticles in different inert ambiences. The particles were found to have a mean size between 30 nm to 45 nm. The size and shape of the particles generated was found be strongly affected by the ambient medium pressure (Sarathi et al., 2007). Nanopowders of aluminium and aluminium oxides are produced using an arc operated between a refractory rod anode and a hollow cathode. The powder consists of irregular spherical particles with size less than 100 nm (Haidar, 2009). Polyhedral aluminium nanoparticles are synthesized using the aerosol route. The size of aluminium particles lies in the range of 50 nm to 100 nm with an average particle size of approximately 87 nm (Kalpowitz et al., 2010). A chemical synthesis method was used to produce nano sized alumina particles. The average particle size was found to be 43 nm (Hemalatha et al., 2011). Aluminium nanoparticles are synthesized in dibutyl ether using a wet chemical process. The mean size of the particles was found to be in the range of 139 nm to 614 nm. The mean size of aluminium particles was reduced to approximately 35 nm when oleic acid was used as a surfactant (Lee and Kim, 2011). The principle of laser ablation was used to produce aluminium nanoparticles in water and ethanol saturated with hydrogen. The particles were found to be spherical in shape and the size lies in the range of 30 nm to 50 nm (Kuzmin et al., 2012). The aluminium nanoparticles are produced under atmospheric pressure using the plasma processing route. The nanoparticles are observed to be mostly spherical in shape and the size lies in the range of 10 nm to 140 nm (Mandilas et al., 2013). The colloidal nanoparticles of aluminium in different liquid environments—distilled water, ethanol, and acetone—are synthesized using the pulsed laser ablation method. The mean size of the particles in distilled water, ethanol, and acetone was found to be 58 nm, 22 nm, and 35 nm. The liquid medium was found to have a strong effect on the size of the particles produced (Mahdieh and Fattahi, 2015). The silver nanoparticles are produced by exploding silver wire (300 μm diameter) in liquid media. The mean size of the silver particles was found to be 25 nm (Cho et al., 2007).

1.4.2 Review on Nanoparticles Stabilization and Application Characteristics

Researchers have carried out studies on the dispersion of micron or millimeter sized solid particles in a liquid seeking to point out the mechanism that provides useful information about ultrasonic velocity related to particle size, concentration, and mechanical properties of the constituents (Kytomaa, 1995; Arenas et al., 2002). The concentration characterization of micron-sized particles dispersed in water was studied and it was found that the colloidal suspension of micron particles reduces ultrasonic velocity. The Urick's equation relating the ultrasonic velocity to slurry concentration was used for applications in which micro particles are involved (Arenas et al., 2002). The polymer colloidal suspensions of chosen concentrations of 0.1–2.0 wt% gold (Au) and copper (Cu) nanoparticles are synthesized using chemical routes. The ultrasonic velocity measurement was made in PVA polymer nano-colloids of Au and Cu at 2 MHz frequency using interferometric and pulse-echo techniques. The results showed that for 0.2 wt% Au-PVA colloid samples, the ultrasonic velocity was found to be a minimum value of 1496 m/s at 30°C. For a 0.5 wt% Cu-PVA sample the ultrasonic velocity appears to be a minimum value of 1501 m/s (Yadav et al., 2008). The concentration characterization of DI water based titanium dioxide (TiO_2) and alumina (Al_2O_3) nanofluids using the ultrasound technique was reported.

The TiO_2 and Al_2O_3 nanofluids is prepared by dispersing TiO_2 (size range: 25 nm to 70 nm) and Al_2O_3 nanoparticles (size range: 20 nm to 50 nm) of the desired amount in DI water using an ultrasonic stirrer. The Urick's equation relating the ultrasound velocity to nanofluids concentrations of 0.1, 0.5, 1, and 2 wt% holds good. The nanoparticles change the ultrasonic velocity by influencing the acoustic time of flight (Chakraborty et al., 2011). The ultrasound velocity in different known concentrations (0.05, 0.1, 0.15, and 0.2 vol%) of water based TiO_2 and Al_2O_3 nanofluids prepared by ultrasonication was studied. The measurement has been made at 3.5 MHz of frequency using the pulse-echo technique. The nanoparticles increase the ultrasonic velocity for some concentrations and were found to be maximum, i.e. 1530 m/s for 0.1 vol% TiO_2-water nanofluid and 1528 m/s for 0.15 vol% Al_2O_3-water nanofluid which is higher than the velocity in the base fluid (1497 m/s) (Tajik et al., 2012).

Improving thermal conductivity of the base fluid using nanoparticles is currently one of the major research fields in the area of nanoparticle technology. Several researchers have worked out how to explain the increase in thermal conductivity of nanofluids in various ways. However, within the domain of thermal science, nanofluids were originally researched for their unaccountable large effective thermal conductivities, which gave rise to the idea of exploiting nanofluids as heat transfer fluids. As a result, nanofluid thermal conductivity and viscosity are all of interest. Viscosity is also an important parameter for any application of colloids. A number of measurements have been reported in the literature and the general conclusion appears to be that nanoparticle aggregation and/or nanoparticle size has a large impact on the potential enhancement of viscosity (Taylor et al., 2013).

Dispersion and stability are the essential characteristics in the enhancement of the thermal conductivity of nanofluids. Uniform dispersion and stable suspension of nanofluids are key to most applications of nanofluids since the final properties of nanofluids such as thermal conductivity and viscosity are determined by the quality of the dispersed state of the suspension (Jang et al., 2007; Hong and Yang, 2001; Wang et al., 2009). The thermal conductivity data of nanofluids varies widely and recent reviews have tried to assess the large amount of data to help in understanding the mechanism of thermal conductivity enhancement of nanofluids. Mechanisms proposed to explain thermal conductivity enhancement include Brownian motion of nanoparticles, layering of fluid around nanoparticles, and near field radiative heat transfer (Prasher et al., 2006; Xue et al., 2004; Keblinski et al., 2002; Abdallah, 2006).

The effect of surface area/volume ratio on thermal conductivity of ethylene glycol based copper nanofluid was studied and demonstrates that the thermal conductivity of the fluid is enhanced due to the substantial impact of a specific area (Vadasz, 2006). For thermal conductivity enhancement of nanofluids, Brownian motion of nanoparticles is the chief mechanism that controls their thermal behavior (Jang and Choi, 2004). The ethylene glycol based copper nanofluid synthesized by vapor synthesis technique and the thermal conductivity was found to have improved by nearly 40% through the dispersion of 0.3 vol% nanofluid concentration (Eastman et al., 2001). Copper-ethylene glycol with polyvinylpyrrolidone (surfactant) nanofluid is prepared by reducing a mixture of copper sulfate pentahydrate in ethylene glycol with sodium hypophosphite monohydrate and it has been observed that the thermal conductivity was enhanced by almost 9% with 0.1 vol% concentration of nanofluid (Zhu et al., 2004a). The copper-water nanofluid using the chemical reduction process was prepared and it was observed that thermal conductivity enhancement of water was found to be about 23.8% with 0.1 vol% concentration. The enhancement in heat transfer of copper nanofluid was due to the suspension of nanoparticles (Liu et al., 2006) in the base fluid and the disordered movement of very fine particles which hasten the energy swap process in the fluid (Xuan and Roetzel, 2000). The thermal conductivity and viscosity of ethylene glycol based copper nanofluid have been studied and it was found that the increase in measured thermal conductivity value as compared to the predicted Maxwell principle was doubled (Garg et al., 2008). The thermal conductivity of EG was found to have increased by 25%–70% and 11%–33% by dispersing copper and iron nanoparticles in EG, respectively, with a change in nanoparticle volume fraction. Also, the thermal conductivity of copper nanofluids are observed to be superior compared to iron nanofluids, for a given particle concentration (Sinha et al., 2009). The Critical Heat Flux (CHF) of copper nanofluids prepared in both distilled water and distilled water with sodium lauryl sulfate surfactant was studied. Compared to pure water, the CHF of copper-water nanofluids was found to have increased with increased particles concentration, but this trend was found to be the opposite in the case of copper-water with stabilizer nanofluids. The decrease in CHF was attributed to the reduction in surface tension of water by surfactant and high bubble dynamic (Kathiravan et al., 2010). The copper-water and copper-water with laurate salt (surfactant) nanofluids are prepared by dispersing copper nanoparticles into DI water. The sedimentation of copper nanoparticles was found to be reduced using laurate salt with water and a significant increase in the thermal conductivity of the fluid was observed (Xuan and Li, 2000). Colloids of copper particles were used to improve the cooling rate of automobile engines. Such improvement can be used to remove engine heat with a reduced size and weight of a radiator (cooling system). This results in the reduction of aerodynamic drag, fluid pumping, and fan requirements, which in turn leads to less fuel consumption (Kumar et al., 2009; Ollivier et al., 2006; Leong et al., 2010).

The copper oxide nanofluids was produced by dispersing commercially available CuO nanoparticles of average size 33 nm in base fluids—DI water, EG, and engine oil. The stability of nanofluids was found to be strongly affected by the characteristics of the suspended particles and base fluids. The thermal conductivity of nanofluids increased compared to base fluids with increased volume fraction of nanoparticle (Hwang et al., 2007). The propylene glycol based alumina (Al_2O_3) and titania (TiO_2) nanofluids of an average size of 89 nm and 115 nm were prepared with good stable dispersion using the two-step method by prolonged ultrasonication without surfactants. The thermal conductivity of nanofluids increases non-linearly with particle concentration. The enhancement of thermal conductivity was temperature independent and parallels the behavior of pure propylene glycol (Palabiyik et al., 2011). The key features such as ballistic, rather than disseminate, and the heat transfer nature in the nanoparticles affect the thermal characteristics of the nanofluids (Keblinski et al., 2002). Many of the reported anomalous enhancements in thermal conductivities of nanofluids could not be produced again (Keblinski et al., 2005). The alumina nanoparticles added into DI water and EG have led to increased thermal conductivities of the suspensions (0.741 W/m K for alumina-DI water nanofluid and 0.326 W/m K for alumina-EG nanofluid) compared to the base fluids (0.613 W/m K—DI water and 0.252 W/m K—EG) at a volume fraction of 0.05. This was attributed to high specific surface area of nanoparticle (Xie et al., 2002). The thermal performance of the heat pipe using commercially available copper nanoparticles of size 40 nm dispersed in aqueous solution of n-Butanol was analyzed. The thermal conductivity of the heat pipe was enhanced by using aqueous n-Butanol based copper nanofluid compared to only DI water and copper-DI water nanofluid (Senthilkumar et al., 2010). The increase in thermal conductivity of nanofluids viz. Al_2O_3, MgO, SiO_2, and Fe_2O_3 nanofluids could be affected by multi-faceted factors including the volume fraction of the dispersed nanoparticles, the tested temperature, thermal conductivity of the base fluid, size of the dispersed nanoparticles, pretreatment process, and the additives of the fluids (Xie et al., 2011). The Al_2O_3 nanofluids were prepared by dispersing commercially available spherical Al_2O_3 nanoparticles of average size of about 13 nm in DI water + sodium dodecyl sulfate surfactant mixture using a two-step method. Uniform and improved stable dispersion of Al_2O_3/DI water nanofluids were found to be achieved with the optimal concentration of a surfactant. The highest thermal conductivity ratio (1.07) was achieved at an optimal surfactant concentration of 0.5 mass fractions (%) and particle concentration of 0.5 volume fraction (%) (Xia et al., 2014). The thermal conductivity of alumina and copper oxide nanoparticles of mean size 28 nm and 23 nm, respectively, suspended in DI water, EG, and engine oil was measured using the steady state parallel plate method. The thermal conductivities of nanofluids were found to have increased compared to the base fluids (Wang and Xu, 1999). The alumina particles play an important role in the performance of solid rocket motors. They cause two-phase losses of specific impulse (produce a negative effect on the total energy/dynamics/motion of the solid motor). The particles also provide damping for the acoustic oscillations in the combustion chamber (Hermsen, 1981). The role of nanometric and micrometric aluminium particles in various explosives was investigated and it was concluded that the mixture of TNT (Trinitrotoluene—a pale yellow, solid organic nitrogen compound used chiefly as an explosive) and nano-aluminium demonstrated higher detonation velocities and heats of detonation. This is due to nano-aluminium reacting faster than regular micron-sized particles in TNT and aluminium compositions (Brousseau and Anderson, 2002). The copper nanoparticles and acoustic cavitation have a great and significant effect on heat transfer in the fluid. A single-phase convection heat transfer was found to have increased with respect to pool boiling of the nanofluids. This was due to the addition of a small amount of copper nanoparticles, while boiling heat transfer was reduced. But, with cavitation in the presence of acoustic field produced in the working fluid, the heat transfer was found to have increased by copper nanoparticles, irrespectively of heat flux (Zhou, 2004). The use of nanofluids in a commercial heat exchanger, in addition to the physical properties, the type of flow (laminar or turbulent) inside the heat exchanging equipment should play an important role in the effectiveness of a nanofluid (Saidur et al., 2011). In the case of heat exchanging equipment with increased thermal duties, where volume is also a matter, and especially in laminar flow conditions, the use of a nanofluid instead of a conventional fluid was found to be advantageous (Pantzali et al., 2009; Saidur et al., 2011). The use of aluminium nanoparticles as an ingredient for solid rocket propellant and explosive formulation increases the burning rate, and results in higher and faster energy release (Sindhu et al., 2007; Kassaee and Buazar, 2009; Rai et al., 2006). It has been shown that when the propellant burning rate increases by adding aluminium nanoparticles, this results in sub-micron aluminium oxide as a combustion product. This could be effective in decreased smoke applications because the smoke visibility varies as the square of the product-oxide. Thus, the use of aluminium nanoparticles allows reduced smoke while still ensuring the high specific impulse offered by aluminized propellants (Ivanov et al., 2003; De Luca et al., 2005).

1.5 MOTIVATION

The research on nanoparticle technology has been growing fast, as witnessed by the number of researchers working in this domain. Nanoparticles have found a wide range of applications in industry, the scientific and medical fields because of their novel properties. The review of nanoparticles research as discussed in Section 1.4 reveals that many methods have been established to synthesize and characterize nanoparticles, but nanoparticles synthesis using a corona discharge micromachining—Electrical Discharge Micromachining (EDMM)—has yet to be addressed in detail. To provide adequate information on the chemical and physical aspects of the established synthesis methods of nanoparticles, diagnostic tools for characterization, and their promising applications, and also to facilitate the widespread information of the newly explored EDMM technology used for synthesis of nanoparticles and their characterization, has motivated us to write this book. The book attempts to present important selected topics in such a manner that it is sufficiently readable and thorough so that it can reach a wide audience of those who might not have a background in these areas. Moreover, people working in one area can understand developments in other areas. The book is primarily focused on the EDMM which is a simple, compact, versatile, and cost-effective method that can synthesize nanoparticles of conductive engineering materials in a single-step process. By closely controlling the EDMM process parameters namely voltage, current, frequency, duty cycle, and pulse duration, it is possible to synthesize a large quantity of nanoparticles of required size and morphology. In this book, detailed information on the synthesis of copper and aluminium nanoparticles using the indigenously developed EDMM setup and size, shape and distribution, concentration, thermal conductivity, viscosity, and optical absorption characterization of the colloidal suspension of particles is provided.

1.6 METHODOLOGY

This book has focused on the introduction to nanoparticles, their classification, potential applications, and exploration of research so that readers can obtain an appreciation of developments outside their current knowledge. Nanoparticles of a size less than 100 nm show new chemistry and physics, leading to new behavior. The dependence of the behavior on particle sizes could allow one to engineer their properties. A wide variety of established methods for synthesis and characterization of nanoparticles are explained briefly in the book for the systematic and coherent realization of the readers.

The book focused on the synthesis of nanoparticles using a corona discharge micromachining—Electrical Discharge Micromachining (EDMM)—method. To date the EDMM method has been used for machining of micro-features like micro holes and micro channels on various engineering conducting and semiconducting materials with required accuracy and precision. These features are obtained by removing a tiny bit of material in the form of debris using a repetitive corona discharge (spark discharge) which occurs between the tool electrode (cathode) and the workpiece electrode (anode). But, attention has not been paid to the debris removed from the electrode materials during the machining operation as nanoparticles. So, in this book, an innovative approach—development and fabrication of an indigenous EDMM prototype system coupled with a piezoactuated tool feed system and ultrasonicator—was attempted to synthesize the nanoparticles. The developed EDMM method is considered as a single-step method used for the synthesis of nanoparticles in combination with the production of colloidal suspensions at high yield.

In this subject, experiments have been carried out on copper and aluminium plates as workpieces and wires as tool materials to synthesize the respective copper and aluminium nanoparticles in the dielectric medium. The structural, chemical, and application characterization of synthesized colloidal suspensions of copper and aluminium particles are carried out using the various diagnostic studies, namely Transmission Electron Microscopy (TEM), Energy Dispersive Analysis by X-Rays (EDAX), Selected Area Electron Diffraction (SAED), ultraviolet-visible spectroscopy, ultrasound velocity method, thermal conductivity measurement, and viscosity measurement methods, and the results are presented. The effect of different stabilizers that influence the interaction between the nucleated nanoparticles in regards to size, shape, and distribution are discussed in the book. The heat transfer capability of the generated colloidal suspensions of copper and aluminium nanoparticles are studied using the developed thermal management setup.

REFERENCES

Abdallah, P.B. (2006) Heat transfer through near-field interactions in nanofluids. *Applied Physics Letters*, 89, 113117/1–113117/3.

Abreo, A.F.V. *Preparation of Copper-Bearing Nanofluids for Thermal Applications*, USA: Proquest LLC, 2009.

Arenas, T.E.G., Segura, L.E. and Sarabia, E.R.F. (2002) Characterization of suspensions of particles in water by an ultrasonic resonant cell. *Ultrasonics*, 39, 715–727.

Basha, J.S. and Anand, R.B. (2011) An experimental study in a CI engine using nanoadditive blended water-diesel emulsion fuel. *International Journal of Green Energy*, 8 (3), 332–348.

Brousseau, P. and Anderson, C. (2002) Nanometric aluminum in explosives. *Propellants, Explosives, Pyrotechnics*, 27, 300–306.

Chakraborty, S., Saha, S.K., Pandey, J.C. and Das, S. (2011) Experimental characterization of concentration of nanofluid by ultrasonic technique. *Powder Technology*, 210, 304–307.

Chang, H., Wu, Y.C., Chen, X.Q. and Kao, M.J. *Fabrication of Cu Based Nanofluid with Superior Dispersion*, Taiwan: PhD Thesis, National Taipei University of Technology, 2000.

Cho, C.H., Park, S.H., Choi, Y.W. and Kim, B.G. (2007) Production of nanopowder by wire explosion in liquid media. *Surface and Coatings Technology*, 201, 4847–4849.

De Luca, L.T., Galfetti, L., Severini, F., Meda, L., Marra, G., Vorozhtsov, A.B., Sedoi, V.S. and Babuk, V.A. (2005) Burning of nano-aluminized composite rocket propellants. *Combustion, Explosion, and Shock Waves*, 41, 680–692.

Dokhan, A., Price, E., Seitzman, J. and Sigman, R. (2002) The effects of bimodal aluminium with ultra-fine aluminium on the burning rates of solid propellants. *Proceedings of Combustion Institute*, 29, 2939–2945.

Drelich, J., Li, B., Bowen, P., Hwang, J., Mills, O. and Hoffman, D. (2011) Vermiculite decorated with copper nanoparticles: antibacterial hybrid material. *Applied Surface Science*, 257, 9435–9443.

Eastman, J.A., Choi, S.U.S., Li, S., Yu, W. and Thompson, L.J. (2001) Anomalously increased effective thermal conductivities of ethylene glycol-based nanofluids containing copper nano-particles. *Applied Physics Letters*, 78, 718–720.

Gadhe, J.B. and Gupta, R.B. (2007) Hydrogen production by methanol reforming in supercritical water: catalysis by in-situ-generated copper nanoparticles. *International Journal of Hydrogen Energy*, 32, 2374–2381.

Garg, J., Poudel, B., Chiesa, M., Gordon, J.B., Ma, J.J., Wang, J.B., Ren, Z.F., Kang, Y.T., Ohtani, H., Nanda, J., McKinley, G.H. and Chen, G. (2008) Enhanced thermal conductivity and viscosity of copper nanoparticles in ethylene glycol nanofluid. *Journal of Applied Physics*, 103, 074301/1–074301/6.

Gleiter, H. (2000) Nanostructured materials: basic concepts and microstructure. *Acta Materialia*, 48 (1), 1–20.

Haidar, J. (2009) Synthesis of Al nanopowders in an anodic arc. *Plasma Chemistry and Plasma Processing*, 29, 307–319.

Hemalatha, J., Prabhakaran, T. and Nalini, R.P. (2011) A comparative study on particle-fluid interactions in micro and nanofluids of aluminium oxide. *Microfluid Nanofluid*, 10, 263–270.

Hermsen, R.W. (1981) Aluminum oxide particle size for solid rocket motor performance prediction. *Journal of Spacecraft and Rockets*, 18 (6), 483–490.

Hong, T.K. and Yang, H.S. (2001) Nanoparticle dispersion dependent thermal conductivity in nanofluids. *Journal of the Korean Physical Society*, 47, 321–324.

Hosokawa, M., Nogi, K., Naito, M. and Yokoyama, T. *Nanoparticle Technology Handbook*, The Netherlands: Elseveir Science, 2012.

Hwang, Y., Lee, J.K., Lee, C.H., Jung, Y.M., Cheong, S.I., Lee, C.G., Ku, B.C. and Jang, S.P. (2007) Stability and thermal conductivity characteristics of nanofluids. *Thermochimica Acta*, 455, 70–74.

Ivanov, Y.F., Osmonoliev, M.N., Sedoi, V.S., Arkhipov, V.A., Bondarchuk, S.S., Vorozhtsov, A.B., Korotkikh, A.G. and Kuznetsov, V.T. (2003) Production of ultra-fine powders and their use in energetic compositions. *Propellants, Explosives, Pyrotechnics*, 28, 319–333.

Jang, S.P. and Choi, S.U.S. (2004) Role of Brownian motion in the enhanced thermal conductivity of nanofluids. *Applied Physics Letters*, 84, 4316–4318.

Jang, S.P., Hwang, K.S., Lee, J.H., Kim, J.H., Lee, B.H. and Choi, S.U.S. (2007) Effective thermal conductivities and viscosities of water-based nanofluids containing Al_2O_3 with low concentration. *Proceedings of 7th IEEE Conference on Nanotechnology*, Hong Kong, 1011–1014.

Johnston, R.L. and Wilcoxon, J.P. *Frontiers of Nanoscience*, Volume 3, The Netherlands: Elsevier Science, 2012.

Kalpowitz, D.A., Jouet, R.J. and Zachariah, M.R. (2010) Aerosol synthesis and reactive behavior of faceted aluminum nanocrystals. *Journal of Crystal Growth*, 312, 3625–3630.

Karasev, V.V., Onischuk, A.A., Glotov, O.G., Baklanov, A.M., Maryasov, A.G., Zarko, V.E., Panfilov, V.N., Levykin, A.I. and Sabelfeld K.K. (2004) Formation of charged aggregates of Al_2O_3 nanoparticles by combustion of aluminum droplets in air. *Combustion and Flame*, 138, 40–54.

Kassaee, M.Z and Buazar, F. (2009) Al nanoparticles: impact of media and current on the arc fabrication. *Journal of Manufacturing Processes*, 11, 31–37.

Kathiravan, R., Kumar, R., Gupta, A. and Chandra, R. (2010) Preparation and pool boiling characteristics of copper nanofluids over a flat plate heater. *International Journal of Heat and Mass Transfer*, 53, 1673–1681.

Keblinski, P., Eastman, J. and Cahill, D. (2005) Nanofluid for thermal transport. *Materials Today*, 8, 36–44.

Keblinski, P., Phillpot, S.R., Choi, S.U.S. and Eastman, J.A. (2002) Mechanisms of heat flow in suspensions of nano-sized particles (nanofluids). *International Journal of Heat and Mass Transfer*, 45, 855–863.

Kristal, M., Ban, I. and Gyergyek, S. (2013) Preparation of nanosized copper and cadmium chalcogenides by mechanochemical synthesis. *Materials and Manufacturing Processes*, 28 (9), 1009–1013.

Kulkarni, D., Vajjha, R., Das, D. and Oliva, D. (2008) Application of aluminum oxide nanofluids in diesel electric generator as jacket water coolant. *Applied Thermal Engineering*, 28, 1774–1781.

Kumar, S., Singh, R., Singh, T.P. and Sethi, B.L. (2009) Surface modification by electrical discharge machining: a review. *Journal of Materials Processing Technology*, 209 (8), 3675–3687.

Kuzmin, P., Shafeev, G., Viau, G., Fonrose, B., Barberoglou, M., Stratakis, E. and Fotakis, C. (2012) Porous nanoparticles of Al and Ti generated by laser ablation in liquids. *Applied Surface Science*, 258, 9283–9287.

Kytomaa, H.K. (1995) Theory of sound propagation in suspensions: a guide to particle size and concentration characterization. *Powder Technology*, 82, 115–121.

Lee, H.M. and Kim, Y.J. (2011) Preparation of size-controlled fine Al particles for application to rear electrode of Si solar cells. *Solar Energy Materials and Solar Cells*, 95, 3352–3358.

Lee, Y.S., Bora, B., Yap, S.L. and Wong, C.S. (2012) Effect of ambient air pressure on synthesis of copper and copper oxide nanoparticles by wire explosion process. *Current Applied Physics*, 12, 199–203.

Leong, K.Y., Saidur, R., Kazi, S.N. and Mamun, A.H. (2010) Performance investigation of an automotive car radiator operated with nanofluid based coolants (nanofluid as a coolant in a radiator). *Applied Thermal Engineering*, 30 (17), 2685–2692.

Liu, M., Lin, M.C., Tsai, C.Y. and Wang, C. (2006) Enhancement of thermal conductivity with copper for nanofluids using chemical reduction method. *International Journal of Heat and Mass Transfer*, 49, 3028–3033.

Lo, C.H., Tsung, T.T, Chen, L.C., Su, C.H. and Lin, H.M. (2005) Fabrication of copper oxide nanofluid using submerged arc nanoparticle synthesis system (SANSS). *Journal of Nanoparticle Research*, 7, 313–320.

Luther, W. (2004) *Industrial Application of Nanomaterials*, Germany: Future Technology Division.

Mahdieh, M.H. and Fattahi, B. (2015) Size properties of colloidal nanoparticles produced by nanosecond pulsed laser ablation and studying the effects of liquid medium and laser fluence. *Applied Surface Science*, 329, 47–57.

Mandilas, C., Daskalos, E., Karagiannakis, G. and Konstandopoulos, A. (2013) Synthesis of aluminium nanoparticles by arc plasma spray under atmospheric pressure. *Materials Science and Engineering B*, 178, 22–30.

Ollivier, E., Bellettre, J., Tazerout, M. and Roy, G.C. (2006) Detection of knock occurrence in a gas SI engine from a heat transfer analysis. *Energy Conversion and Management*, 47, 879–893.

Palabiyik, I., Musina, Z., Witharana, S. and Ding, Y. (2011) Dispersion stability and thermal conductivity of propylene glycol-based nanofluids. *Journal of Nanoparticle Research*, 13, 5049–5055.

Pantzali, M.N., Mouza, A.A. and Paras, S.V. (2009) Investigating the efficacy of nanofluids as coolants in plate heat exchangers (PHE). *Chemical Engineering Science*, 64, 3290–3300.

Pfeifer, P., Schubert, K. and Emig, G. (2005) Preparation of copper catalyst washcoats for methanol steam reforming in microchannels based on nanoparticles. *Applied Catalysis A: General*, 286, 175–185.

Ponce, A.A. and Klabunde, K.J. (2005) Chemical and catalytic activity of copper nanoparticles prepared via metal vapor synthesis. *Journal of Molecular Catalysis A: Chemical*, 225, 1–6.

Poole, C.P. and Owens, F.J. *Introduction to Nanotechnology*, USA: Wiley Interscience, 2003.

Prasher, R., Bhattacharya, P. and Phelan, P.E. (2006) Brownian-motion-based convective-conductive model for the effective thermal conductivity of nanofluids. *Journal of Heat Transfer*, 128 (6), 588–595.

Rai, A., Park, K., Zhou, L. and Zachariah M.R. (2006) Understanding the mechanism of aluminium nanoparticle oxidation. *Combustion Theory and Modelling*, 10 (5), 843–859.

Raja, M., Subha, J., Ali, F.B. and Ryu, S.H. (2008) Synthesis of copper nanoparticles by electro-reduction process. *Materials and Manufacturing Processes*, 23, 782–785.

Robertis, E.D., Cosme, E.H.H., Neves, R.S., Kuznetsov, A.Y., Campos, A.P.C., Landi, S.M. and Achete, C.A. (2012) Application of the modulated temperature differential scanning calorimetry technique for the determination of the specific heat of copper nanofluids. *Applied Thermal Engineering*, 41, 10–17.

Saidur, R., Leong, K.Y. and Mohammad, H.A. (2011) A review on applications and challenges of nanofluids. *Renewable and Sustainable Energy Reviews*, 15, 1646–1668.

Sarathi, R., Sindhu, T.K. and Chakravarthy, S.R. (2007) Generation of nano aluminium powder through wire explosion process and its characterization. *Materials Characterization*, 58 (2), 148–155.

Senthilkumar, R., Vaidyanathan, S. and Sivaraman, B. (2010) Performance analysis of heat pipe using copper nanofluid with aqueous solution of n-butanol. *International Journal of Mechanical and Materials Engineering*, 1 (4), 251–256.

Sindhu, T.K., Sarathi, R. and Chakravarthy, S.R. (2007) Generation and characterization of nano aluminium powder obtained through wire explosion process. *Bulletin of Materials Science*, 30 (2), 187–195.

Sinha, K., Kavlicogulu, B., Liu, Y., Gordaninejad, F. and Graeve, O.A. (2009) A comparative study of thermal behavior of iron and copper nanofluids. *Journal of Applied Physics*, 106, 064307/1-064307/7.

Suyal, G. (2003) Bimetallic colloids of silver and copper in thin films: sol-gel synthesis and characterization. *Thin Solid Films*, 426, 53–61.

Tajik, B., Abbassi, A., Avval, M.S. and Najafabadi, M.A. (2012) Ultrasonic properties of suspensions of TiO_2 and Al_2O_3 nanoparticles in water. *Powder Technology*, 217, 171–176.

Taylor, R., Coulombe, S., Otanicar, T., Phelan, P., Gunawan, A., Lv, W., Rosengarten, G., Prasher, R. and Tyagi, H. (2013) Small particles, big impacts: a review of the diverse applications of nanofluids. *Journal of Applied Physics*, 113, 011301/1-011301/19.

Vadasz, P. (2006) Heat conduction in nanofluid suspensions. *Journal of Heat Transfer*, 128, 465–477.

Wang, X. and Xu, X. (1999) Thermal conductivity of nanoparticle-fluid mixture. *Journal of Thermophysics and Heat Transfer*, 13 (4), 474–480.

Wang, X.J., Li, X. and Yang, S. (2009) Influence of pH and SDBS on the stability and thermal conductivity of nanofluids. *Energy and Fuels*, 23, 2684–2689.

Wen, J., Li, J., Liu, S. and Chen, Q. (2011) Preparation of copper nanoparticles in a water/oleic acid mixed solvent via two-step reduction method. *Colloids and Surfaces A: Physicochemical and Engineering Aspects*, 373, 29–35.

Xia, G., Jiang, H., Liu, R. and Zhai, Y. (2014) Effects of surfactant on the stability and thermal conductivity of Al_2O_3/de-ionized water nanofluids. *International Journal of Thermal Sciences*, 84, 118–124.

Xie, H., Wang, J., Xi, T., Liu, Y., Ai, F. and Wu, Q. (2002) Thermal conductivity enhancement of suspensions containing nanosized alumina particles. *Journal of Applied Physics*, 91, 4568–4572.

Xie, H., Yu, W., Li, Y. and Chen, L. (2011) Discussion on the thermal conductivity enhancement of nanofluids. *Nanoscale Research Letters*, 6, 124/1–124/12.

Xuan, Y. and Li, Q. (2000) Heat transfer enhancement of nanofluids. *International Journal of Heat and Fluid Flow*, 21, 58–64.

Xuan, Y.M. and Roetzel, W. (2000) Conceptions for heat transfer correlation of nanofluids. *International Journal of Heat and Mass Transfer*, 43, 3701–3707.

Xue, L., Keblinski, P., Phillpot, S.R., Choi, S.U.S. and Eastman, J.A. (2004) Effect of liquid layering at the liquid–solid interface on thermal transport. *International Journal of Heat and Mass Transfer*, 47, 4277–4284.

Yadav, R.R., Mishra, G., Yadawa, P.K., Kor, S.K., Gupta, A.K., Raj, B. and Jayakumar, T. (2008) Ultrasonic properties of nanoparticles-liquid suspensions. *Ultrasonics*, 48, 591–593.

Zhang, Y., Jiang, W. and Wang, L. (2010) Microfluidic synthesis of copper nanofluids. *Microfluid Nanofluid*, 9, 727–735.

Zhou, D.W. (2004) Heat transfer enhancement of copper nanofluid with acoustic cavitation. *International Journal of Heat and Mass Transfer*, 47, 3109–3117.

Zhu, H., Lin, Y. and Yin, Y. (2004a) A novel one-step chemical method for preparation of copper nanofluids. *Journal of Colloid and Interface Science*, 277, 100–103.

Zhu, H., Zhang, C. and Yin, Y. (2004b) Rapid synthesis of copper nanoparticles by sodium hypophosphite reduction in ethylene glycol under microwave irradiation. *Journal of Crystal Growth*, 270, 722–728.

CHAPTER 2

Synthesis Methods

2.1 INTRODUCTION

Synthesis of nanoparticles is the first cornerstone nanotechnology. The exploration on the novel physical properties and potential applications of nanoparticles is possible only when nanoparticles are synthesized with the desired size, shape and distribution, and chemical composition. The work on the synthesis of nanoparticles has increased considerably in the last two decades when nanotechnology emerged as a new scientific field. A wide variety of methods has been developed for the synthesis of nanoparticles of definite size and morphology. Therefore, in this chapter we are concerned with the discussion of the various established synthesis methods of nanoparticles in brief. Further, to realize the formulation of colloids from a general perspective, the steps for colloids formation are briefly discussed in this chapter.

2.2 SYNTHESIS METHODS

Two basic approaches have been developed and used for the synthesis of nanoparticles: top-down and bottom-up. In the top-down approach, nanoparticles are synthesized by mechanical crushing of the bulk material, whereas in the bottom-up approach, nanoparticles are synthesized nanoparticles generated from the atomic or molecular level by growth and assembly either atom-by-atom, molecule-by-molecule, or cluster-by-cluster (Figure 2.1).

The top-down approach refers to the mechanical based nanoparticles synthesis methods, whereas the bottom-up approach refers to the chemical (liquid phase reaction) and physical (vapor phase reaction) based nanoparticles synthesis methods. Figure 2.2 shows the various methods of nanoparticles synthesis and their subdivisions. These methods are explained briefly in the following section.

2.2.1 Mechanical Methods

2.2.1.1 Ball Milling Method

Ball milling uses the mechanical energy to be applied on the bulk solid materials to break the bonding between atoms or molecules. In this process, small balls are allowed to rotate inside a drum containing bulk structures which break down into crystallite nanoparticles. This is an established technology, which is simple and cheap, but its disadvantages include contamination of the particles resulting from the grinding media, lack of control on the particle size distribution, highly poly-dispersed size distribution,

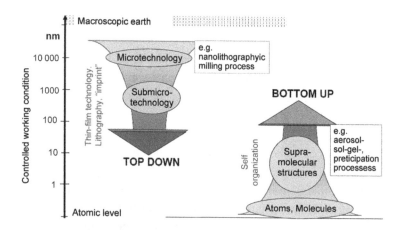

FIGURE 2.1 Top-down and bottom-up approach for nanoparticles synthesis.

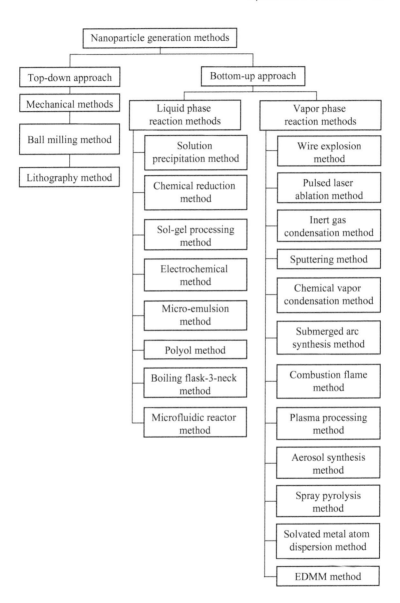

FIGURE 2.2 Various methods of nanoparticles synthesis and their subdivisions.

introduction of internal stresses, partially amorphous state of the powder, and the difficulty in producing the desired particle size and shape. This method is used for generating metal oxides and intermetallic compounds. For example, lithium iron phosphate ($LiFePO_4$) nanoparticles of uniform size distribution of about 50 nm were synthesized by this method from the mixture of iron, lithium phosphate, and iron phosphate powders (Wilson et al., 2002; Xinping et al., 2006).

2.2.1.2 Lithography Method

Lithography is the method of transferring a pattern into a reactive polymer film, termed as resist, which will subsequently be used to replicate that pattern into an underlying thin film or substrate. The feature size less than 100 nm can be obtained by this method. The disadvantages of this method include significant crystallographic damage to the processed patterns, and additional defects may be introduced even during the etching steps. For example, nanowires made by lithography are not smooth and may contain a lot of impurities and structural defects on the surface. Such imperfections would have a significant impact on physical properties and surface chemistry of nanomaterials (Cao, 2004).

2.2.2 Liquid Phase Reaction Methods

2.2.2.1 Solution Precipitation Method

In the precipitation method, an inorganic metal salt (e.g. copper sulphate, copper nitrate, copper acetate, etc.) is dissolved in water, then metal ions form metal hydrate species in the water and these are hydrolyzed by adding a base solution. The hydrolyzed species condense with each other forming metal hydroxide precipitate. This precipitate is filtered out and dried, which is subsequently calcined to obtain the final crystalline nanoparticles. The advantage of this process is that it is economical, but the main drawback is the inability to control the size of the particles and their subsequent aggregation (Gleiter, 1989).

2.2.2.2 Chemical Reduction Method

In the chemical reduction method, nanoparticles are synthesized by the reduction of a suitable precursor salt in an aqueous solution with a reducing agent. This method produces small sized nanoparticles with narrow distribution, but its disadvantages include a complex process and a number of toxic chemicals are used. This method is used for synthesizing metal, non-metal, metal oxide, and alloy nanoparticles (Schmid, 2008; Taylor et al., 2013).

2.2.2.3 Sol-Gel Method

In the sol-gel method, inorganic and organic salts or a metal alkoxides sample is dissolved in water and the hydrolyzed sample is condensed to form nanoparticles. This method involves low processing temperature and molecular level homogeneity, but its disadvantages include costly metallic precursor samples and sensitivity to the atmospheric condition. This method is used for synthesizing metal oxide or metal hydroxide nanoparticles (Gleiter, 1989; Wilson et al., 2002).

2.2.2.4 Electrochemical Method

In the electrochemical method, an electric current is passed between two electrodes—cathode and anode—separated by an electrolyte, and nanoparticles are synthesized by anodic dissolution based on electrolysis phenomenon. This method involves low cost, simple operation, and high flexibility, but its disadvantages include that the chemicals used are toxic and difficult to remove from the nanofluid. This method is used for synthesizing metal or metal oxide nanoparticles (Taylor et al., 2013).

2.2.2.5 Micro-Emulsion Method

In the micro-emulsion method, fine liquid droplets of an organic solvent are dispersed in an aqueous solution containing a metal precursor and reducing agent. The dispersed droplets continuously collide with each other and thus nanoparticles are synthesized through a chemical reaction, nucleation, and growth inside the droplets. This method is simple and suitable for producing mono-dispersed nanoparticles, but its disadvantages include high cost and low yield. This method is used for synthesizing metal, metal oxide, and metal alloy nanoparticles (Cao, 2004).

2.2.2.6 Polyol Method

In the polyol method, nanoparticles are synthesized by heating a suitable organic or inorganic metallic salt in polyol (i.e. alcohol having multiple hydroxyl groups). The polyol acts as a solvent and reducing agent. For example, polyols can effectively reduce metal ions to synthesize nano sized particles of copper, silver, and gold under microwave irradiation, and the approach is referred to as the microwave polyol process. This method is used for synthesizing mono-dispersed and non-agglomerated nanoparticles but its disadvantage is that the solution of metallic salt should be heated to its boiling point and kept under refluxing conditions for a long time. This method is used for synthesizing metal nanoparticles (Figlarz et al., 1985; Fievet et al., 1989; Zhu et al., 2004).

2.2.2.7 Boiling Flask-3-Neck Method

In the boiling flask-3-neck method, the diluted reactants—a solution of metal hydroxide + suitable hydrate—are mixed under a magnetic stirrer at room temperature. The mixed reactant undergoes a chemical reaction yielding a fast reduction to nanoparticles. This method is simple, but its disadvantages include difficulty in controlling the particles size and the size distribution is relatively broad. This method is used for synthesizing metal nanoparticles (Zhang et al., 2010).

2.2.2.8 Microfluidic Reactor Method

In the microfluidic reactor method, the reactants—a solution of metal hydroxide + suitable hydrate + buffer solution—are introduced into the microfluidic reactor by a syringe pump. The mixed reactant which flows through the reaction channel undergoes a chemical change and results in the formation of nanoparticles.

This method eliminates the local variations in the reaction conditions such as concentration and temperature, but its disadvantage includes high cost. This method is used for synthesizing metal nanoparticles (Zhang et al., 2010).

2.2.3 Vapor Phase Reaction Methods

2.2.3.1 Wire Explosion Method

In the wire explosion method, a high voltage is applied to the thin metal wires surrounded by air or inert gases. The energy deposited in a wire by Joule heating melts and evaporates the wire material. Then, condensation and nucleation of wire vapor takes place, which gives rise to the formation of nano sized particles. The nanoparticles were collected on the surface of the membrane filter by evacuating the chamber with the pump. This method controls the size of the nanoparticles and involves high energy efficiency, but its disadvantages include high cost and it is not possible to use for different metals as it is applicable only to high electrical conductivity metals. This method is used for synthesizing metal and metal oxide nanoparticles (Sarathi et al., 2007; Ivanov et al., 1995).

2.2.3.2 Pulsed Laser Ablation Method

In the pulsed laser ablation method, a high power pulsed laser beam is focused on a target material and due to melting and evaporation nanoparticles are synthesized. This method is simple and flexible, and synthesis of nanoparticles can take place in any arbitrary liquid, but its disadvantages include low yield and the mechanism of laser-matter interactions is complex. This method is used for synthesizing metal and metal oxide nanoparticles (Swihart, 2003; Tilaki et al., 2007).

2.2.3.3 Inert Gas Condensation Method

In the inert gas condensation method, metal atoms are evaporated by heating (source: radio frequency/electron beam/resistive heating) the target material in a chamber containing inert gas. The evaporated metal atoms collide with the gas atoms, lose their kinetic energy, and condense to form nanoparticles. This method has the potential to control the particle shape and size, but its disadvantages include low yield and an evaporation problem of low vapor pressure materials. This method is used for synthesizing metal and metal oxide nanoparticles (Gleiter, 1989; Swihart, 2003).

2.2.3.4 Sputtering Method

In the sputtering method, nanoparticles are synthesized from a target material surface by bombardment with high velocity ions of an inert gas produced in an ion gun as a result of the momentum transfer. No melting of the material takes place in this method. This method has the advantage that the composition of the sputtered material generated is the same as that of the target, but its disadvantages include low yield and broad particle size distribution. This method is used for synthesizing metal, metal oxide, and semiconductor nanoparticles (Kathiravan et al., 2010; Suryanarayana and Koch, 2000).

2.2.3.5 Chemical Vapor Condensation Method

In the chemical vapor condensation method, the metal vapor is generated by heating an organometallic precursor and is rapidly condensed in another cold chamber by carrying it with the help of a carrier gas, resulting in the formation of nanoparticles. The nanoparticles are collected by means of filters or electrostatic precipitators. This method produces nanoparticles with narrow size distribution, but its disadvantages include low production rate. This method is used for synthesizing metal and metal oxide nanoparticles (Gleiter, 1989; Swihart, 2003).

2.2.3.6 Submerged Arc Synthesis Method

In the submerged arc synthesis method, the target material is heated by an electric arc and vaporization of the material takes place. The vaporized material condenses to produce nanoparticles in the liquid medium. This method involves easy control over the process parameter, but its disadvantage is low yield. This method is used for synthesizing metal and metal oxide nanoparticles (Chang et al., 2005).

2.2.3.7 Combustion Flame Method

In the combustion flame method, a steady flame is generated by burning a fuel-oxygen mixture. The chemical precursors, (e.g. aluminium and ammonium perchlorate) introduced along with combustibles, experience rapid thermal decomposition in the hot zone of the flame. The continuous stream of nano-clusters produced from the combustion zone is quenched and results in the formation of nanoparticles. This method has the potential to synthesize a large quantity of nanoparticles, but its disadvantages include difficulty in controlling the particle size and the occurrence of aggregation of particles. This method is used for synthesizing metal and metal oxide nanoparticles (Swihart, 2003; Karasev et al., 2004).

2.2.3.8 Plasma Processing Method

In the plasma processing method, the precursor material (e.g. powders of aluminium; iron oxide; zinc oxide; copper oxide; etc.) which is micron-sized particles is injected into a reaction chamber operating at atmospheric pressure and is subjected to arc plasma using graphite electrodes. The plasma causes evaporation of the material leading to synthesis of nanoparticles in the cold wall of the chamber. This method produces a large quantity of nanoparticles at low cost, but its disadvantages include the occurrence of aggregation of particles. This method is used for synthesizing metal, metal oxide, and intermetallic compound nanoparticles (Wilson et al., 2002; Mandilas et al., 2013).

2.2.3.9 Aerosol Synthesis Method

In the aerosol synthesis method, a liquid precursor (e.g. titanium ethoxide; titanium isopropoxide; titanium tetrachloride; tri-iso-butylaluminium) is mistified to make a liquid aerosol (dispersion of uniform droplets of liquid in a gas) and this liquid aerosol solidifies through evaporation of solvent resulting in the synthesis of nanoparticles. Aerosols can be relatively easily produced by sonication or spinning. This method is simple and cheap, but its disadvantages include aggregation, inhomogeneous morphology, and the occurrence of broad distribution of particles. This method is used for synthesizing metal and metal oxide nanoparticles (Cao, 2004; Kalpowitz et al., 2010).

2.2.3.10 Spray Pyrolysis Method

In the spray pyrolysis method, micro-sized droplets of metal salt precursor solution are introduced into a heating furnace, and through evaporation and condensation of metal atoms, nanoparticles are synthesized. In practice, spray pyrolysis involves several steps: (1) producing micro-sized droplets of liquid precursor or precursor solution, (2) evaporation of solvent, (3) condensation of solute, (4) decomposition and reaction of solute, and (5) sintering of solid particles. This method avoids crystal growth and controls the particle size distribution, but its disadvantages include low yield and porous particle is produced in the case of high heating rate or large droplet size. This method is used for generating metal and metal oxide nanoparticles (Messing et al., 1993; Gurav et al., 1993).

2.2.3.11 Solvated Metal Atom Dispersion Method

In the solvated metal atom dispersion method, metal shots are evaporated in a crucible under vacuum and condensed with organic solvent at liquid nitrogen temperature. The organic solvent containing metal atoms are warmed to room temperature resulting in the formation of nanoparticles. This method is used for generating mono-dispersed particles with high yield, but its disadvantage is that the catalytic properties of metal particles are affected by the presence of organic fragments on the particles. This method is used for generating metal, metal oxide, and metal sulfide nanoparticles (Arora and Jagirdar, 2012).

2.2.3.12 *Corona Discharge Micromachining—Electrical Discharge Micromachining (EDMM) Method*

EDMM is an advanced mechanical micromachining method which is based on energy of corona discharge (energy of electrical glow/energy of sudden discharge of electrons, i.e. spark) created between the two electrodes—the tool (cathode) and the workpiece (anode)—surrounded by dielectric fluid. This method is used for machining micro-features like micro holes and micro channels on various engineering conducting materials that are extensively used in a wide variety of applications. These micro-features are obtained by removing a tiny bit of material in the form of debris/chips through melting and evaporation using repetitive spark discharges which appear between the electrodes. Some of the debris is evaporated and some is dispersed in the dielectric fluid. It is quite interesting to note that the debris generated during the EDMM process is very small—micron/nano size—but attention has not been focused on this debris and its applications as nanoparticles in the field of nanotechnology. It is possible to control the size of this debris using controlled process parameters.

2.3 FORMULATION OF COLLOIDS

Materials for nanoparticles and base fluids are different. The nanoparticles are most commonly synthesized in the form of powders. The powdered nanoparticles are dispersed in aqueous or organic liquids for specific applications. Stable colloidal suspensions of nanoparticles are produced by two types of methods: the two-step method and the single-step method. In the two-step method, nanoparticles are first produced and then dispersed in the base fluids. The two-step method includes the liquid chemical reduction method, chemical precipitation, ultrasonic irradiation method, etc. In the single-step method, nanoparticles are generated in base fluids directly and no redispersion process is required. This avoids the process of drying, storage, transportation, and redispersion of nanoparticles which takes place in the two-step method. Thus, the production costs can be reduced. The single-step method is also a preferable way to produce colloids containing high conductivity metals, which allows prevention of oxidation of the particles. The single-step method includes the laser ablation method, sputtering method, microfluidic reactor method, microwave irradiation method, and submerged arc synthesis method.

However, while formulating colloids, both methods suffer from agglomeration of nanoparticles. Therefore, to overcome this issue, in either case, a non-agglomerated and uniformly dispersed colloid is required for successful production of increased properties of colloid and analysis of experimental data. These increased properties of colloid offer several benefits such as higher heat transfer rates, decreased pumping power needs, smaller and lighter cooling systems, electricity savings, improved load carrying capacity, reduced friction, and improved wear resistance. These benefits make colloids a promising application, for example, as coolants and lubricants, in many industries including manufacturing, transportation, energy production, electronics, medicines, etc. (Saidur et al., 2011).

2.4 SUMMARY

In this chapter we discussed the two different approaches for synthesis of nanoparticles. The established mechanical, liquid phase, and vapor phase routes of nanoparticles synthesis were presented. The advantages and disadvantages of the various methods followed by types of nanoparticles synthesized were also discussed. The methods for formulation of colloids were briefly discussed and the promising applications of colloids are outlined.

REFERENCES

Arora, N. and Jagirdar, B.R. (2012) Monodispersity and stability: case of ultrafine aluminium nanoparticles (<5 nm) synthesized by the solvated metal atom dispersion approach. *Journal of Materials Chemistry*, 22, 9058–9063.

Cao, G. *Nanostructures and Nanomaterials: Synthesis, Properties and Applications*, UK: Imperial College Press, 2004.

Chang, H., Tsung, T.T. and Lo, C.H. (2005) A study of nanoparticle manufacturing process using vacuum submerged arc machining with aid of enhanced ultrasonic vibration. *Journal of Materials Science*, 40, 1005–1010.

Fievet, F., Lagier, J.P. and Figlarz, M. (1989) Preparing monodisperse metal powders in micrometer and submicrometer sizes by the polyol process. *Materials Research Bulletin*, 14, 29–34.

Figlarz, M., Fievet, F. and Lagier, J.P. *Process for the Reduction of Metallic Compounds By Polyols, and Metallic Powders Obtained by This Process*, U.S. Patent 4539041, 1985.

Gleiter, H. (1989) Nanocrystalline materials. *Progress in Materials Science*, 33, 223–315.

Gurav, A., Kodas, T., Pluym, T. and Xiong, Y. (1993) Aerosol processing of materials. *Aerosol Science and Technology*, 19 (4), 411–452.

Ivanov, V., Kotov, Y.A., Samatov, O.H., Bohme, R., Karow, H.V. and Schumacher, G. (1995) Synthesis and dynamic compaction of ceramic nanopowders by techniques based on electric pulsed power. *Nanostructured Materials*, 6, 287–290.

Kalpowitz, D.A., Jouet, R.J. and Zachariah, M.R. (2010) Aerosol synthesis and reactive behavior of faceted aluminum nanocrystals. *Journal of Crystal Growth*, 312, 3625–3630.

Karasev, V.V., Onischuk, A.A., Glotov, O.G., Baklanov, A.M., Maryasov, A.G., Zarko, V.E., Panfilov, V.N., Levykin, A.I. and Sabelfeld K.K. (2004) Formation of charged aggregates of Al_2O_3 nanoparticles by combustion of aluminum droplets in air. *Combustion and Flame*, 138, 40–54.

Kathiravan, R., Kumar, R., Gupta, A. and Chandra, R. (2010) Preparation and pool boiling characteristics of copper nanofluids over a flat plate heater. *International Journal of Heat and Mass Transfer*, 53, 1673–1681.

Mandilas, C., Daskalos, E., Karagiannakis, G. and Konstandopoulos, A. (2013) Synthesis of aluminium nanoparticles by arc plasma spray under atmospheric pressure. *Materials Science and Engineering B*, 178, 22–30.

Messing, G.L., Zhang, S.C. and Jayanthi, G.V. (1993) Ceramic powder synthesis by spray pyrolysis. *Journal of The American Chemical Society*, 76, 2707–2726.

Saidur, R., Leong, K.Y. and Mohammad, H.A. (2011) A review on applications and challenges of nanofluids. *Renewable and Sustainable Energy Reviews*, 15, 1646–1668.

Sarathi, R., Sindhu, T.K. and Chakravarthy, S.R. (2007) Generation of nano aluminium powder through wire explosion process and its characterization. *Materials Characterization*, 58 (2), 148–155.

Schmid, G. *Clusters and Colloids: From Theory to Applications*, USA: Wiley and Sons, 2008.

Suryanarayana, C. and Koch, C.C. (2000) Nanocrystalline materials: Current research and future directions. *Hyperfine Interactions*, 130, 5–44.

Swihart, M.T. (2003) Vapor phase synthesis of nanoparticles. *Current Opinion in Colloid and Interface Science*, 8, 127–133.

Taylor, R., Coulombe, S., Otanicar, T., Phelan, P., Gunawan, A., Lv, W., Rosengarten, G., Prasher, R. and Tyagi, H. (2013) Small particles, big impacts: a review of the diverse applications of nanofluids. *Journal of Applied Physics*, 113, 011301/1-011301/19.

Tilaki, R.M., Irajizad, A. and Mahdavi, S.M. (2007) Size, composition and optical properties of copper nanoparticles prepared by laser ablation in liquids. *Applied Physics A: Materials Science and Processing*, 88, 415–419.

Wilson, M., Kannangara, K., Smith, G., Simmons, M. and Raguse, B. *Nanotechnology: Basic Science and Emerging Technologies*, Australia: Chapman and Hall, 2002.

Xinping, A., Hai, L., Xiaoyan, L., Qinlin, L., Bingdong, L. and Hanxi, Y. (2006) One-step ball milling synthesis of $LiFePO_4$ nanoparticles as the cathode material of Li-ion batteries. *Wuhan University Journal of Natural Science*, 11 (3), 687–690.

Zhang, Y., Jiang, W. and Wang, L. (2010) Microfluidic synthesis of copper nanofluids. *Microfluid Nanofluid*, 9, 727–735.

Zhu, H., Zhang, C. and Yin, Y. (2004) Rapid synthesis of copper nanoparticles by sodium hypophosphite reduction in ethylene glycol under microwave irradiation. *Journal of Crystal Growth*, 270, 722–728.

CHAPTER 3

Diagnostic Methods

3.1 INTRODUCTION

Over the last few decades, the nanoparticles in the form of or without colloid have been studied and numerous physical, chemical, thermal, and electrical characteristics of nanoparticles are known. In recent years, one of the critical challenges faced by researchers in the field of nanoparticle technology is the inability and the lack of instruments to examine the individual nanoparticles at atomic level resolution and manipulate them by showing them at the macroscopic level. Also, a better conceptual understanding of various potential application characteristics of nanoparticles increasingly demands the ability and diagnostic tools to characterize the particles. Therefore, it is essential to study the various diagnostic methods that play an important role in the characterization of nanoparticles.

In this chapter, various diagnostic methods that are most commonly used for nanoparticles characterization are discussed. The methods for structural characterization of nanoparticles include: Scanning Electron Microscopy (SEM), Transmission Electron Microscopy (TEM), Energy Dispersive Analysis by X-Rays (EDAX), Selected Area Electron Diffraction (SAED), X-Ray Diffraction (XRD), and Scanning Probe Microscopy (SPM). The methods for chemical characterization include ultraviolet-visible spectroscopy and ionic spectrometry. Then the methods for application characterization include the ultrasound velocity method, thermal and electrical conductivity measurement, and viscosity measurement methods. The discussion in this chapter is focused mainly on the fundamentals and basic principles of diagnostic methods. The technical details and operation procedure are discussed only for diagnostic tools used for application characteristics of nanoparticles.

3.2 DIAGNOSTIC METHODS FOR STRUCTURAL CHARACTERIZATION

3.2.1 Scanning Electron Microscopy (SEM)

SEM is one of the most commonly used methods in characterization of nanoparticles. The resolution of the SEM is of the order of nanometers and can operate at magnifications ranging from 10 to 3 lakhs above. SEM provides information about the size, shape, and chemical composition of a specimen.

In a typical SEM, electrons are produced from the electron gun under vacuum and accelerated towards the target specimen. As the electrons travel towards the specimen, they pass through condenser lenses, which focus the electrons electromagnetically into a beam. The beam then moves through scanning coils, which bend the beam in a raster pattern across the specimen. When the electron beam strikes the specimen, it penetrates the specimen to a certain depth which is proportional to the beam intensity. A number of electrons and photons are emitted from the specimen. But for SEM, only secondary and backscattered electrons are collected. When electrons from the beam come out of the specimen with most of their original energy, they are called backscattered electrons. On the other hand, electrons that are emitted from the specimen with very little energy are called secondary electrons. The energy differentiation between a backscattered electron and a secondary electron is 50 eV. The backscattered and secondary electrons are emitted from the sample in all directions. To obtain as many electrons as possible, SEM has a positively charged collector that draws the electrons to the detector, particularly the low energy secondary electrons. The electron signals that are detected are then amplified and transformed into an image (Cao, 2004; Wang, 1996).

3.2.2 Transmission Electron Microscopy (TEM)

TEM is a technique used for analyzing the size, morphology, crystallographic structure, and composition of a specimen. TEM provides a much higher spatial resolution (above 10 lakhs) than a Scanning Electron Microscope (SEM) and can facilitate the analysis of features nearly at atomic scale (in the range of a few nanometers) using electron beam energies in the range of 60–350 keV.

Unlike the SEM, which relies on dislodged or reflected electrons from the specimen to form an image, the TEM collects the electrons that are transmitted through the specimen. TEM uses an electron gun to produce the primary beam of electrons that are focused by lenses and apertures into a very thin, coherent beam. This beam is then controlled to strike the specimen. A portion of this beam gets transmitted to the other side of the specimen, which is collected and processed to form the image. For crystalline materials, the specimen diffracts the incident electron beam, producing local diffraction intensity variations that can be translated into contrast to form an image. For amorphous materials, contrast is achieved by variations in electron scattering as the electrons traverse the chemical and physical differences within the specimen. The sample preparation is one of the most important tasks during TEM analysis and utmost care has been taken to prepare the samples for TEM study (Das et al., 2008; Yao and Wang, 2005).

3.2.3 Energy Dispersive Analysis by X-Rays (EDAX)

EDAX is a technique used for identifying the elemental composition of the sample. The EDAX analysis system works as an integrated feature of a TEM. During EDAX analysis, an electron beam is bombarded on the sample and the bombarding electrons collide with the sample atoms' own electrons, knocking some of them off in the process. A higher energy electron from an outer shell finally occupies a position vacated by an ejected inner shell electron. However, to be able to do so, the transferring outer electron must give up some of its energy by emitting an X-ray. The amount of energy released by the transmitting electron depends on which shell it is transferring from, as well as which shell it is transferring to. Furthermore, the atom of every element releases X-rays with unique amounts of energy during the transferring process. Thus, by measuring the amounts of energy present in the X-rays being released by a sample during electron beam bombardment, the identity of the atom from which the X-ray was emitted can be measured.

The EDAX spectrum is a plot of how frequently an X-ray is received for each energy level. An EDAX spectrum normally shows peaks corresponding to the energy levels for which the most X-rays had been received. Each of these peaks is unique to an atom, and therefore corresponds to a single element. The higher a peak in the spectrum, the more concentrated the element is in the sample. The sample used for the EDAX study could be same as that used for the TEM analysis.

3.2.4 Selected Area Electron Diffraction (SAED)

SAED is a crystallographic experimental diffraction technique that can be performed using a TEM instrument. SAED can be used to identify crystal structures of the particles in the sample. It is similar to X-ray diffraction, but a unique feature of it is that areas as small as several hundred nanometers in size can be examined, whereas in X-ray diffraction the sample areas are typically of several centimeters in size. An aperture is used to define the area from which a diffraction pattern is formed in a TEM sample. The resulting patterns contain information about the phases present by lattice spacing measurement and sample orientation. The samples used for TEM could also be used for SAED analysis.

3.2.5 X-Ray Diffraction (XRD)

The XRD method is used to get crystallographic information (crystalline phases and orientation of crystals), chemical composition, and physical properties of specimens. When a beam of X-rays strikes a specimen at an angle θ, it is diffracted by the crystalline phases in the specimen according to Bragg's law:

$$n\lambda = 2d \sin\theta \qquad (3.1)$$

where n is the order of scattering, λ is the wavelength of the X-ray, and d is the lattice spacing (spacing between crystal lattice planes).

The intensity of the diffracted X-rays depends on the diffraction angle 2θ and the specimen's orientation. This diffraction pattern is used to identify the specimen's crystalline phases and to determine its structural properties. XRD is a suitable method for characterizing crystalline specimens subjected to uniform elastic strain because it can accurately measure the positions of diffraction peak. When there is a uniform elastic strain then the positions of diffraction peak changes resulting in the change of lattice constants, and thus the change in lattice spacing can be evaluated. In case of specimens subjected to uniform elastic strain, the average crystallite size can be calculated using the Debye–Scherrer formula:

$$D = \frac{k\lambda}{\beta\cos\theta} \qquad (3.2)$$

where D is the average crystallite size, k is the Scherrer's constant (usually 1 or 0.9), λ is the wavelength of X-ray, β is the full width height maximum, and θ is the diffraction angle. The XRD method does not involve any damage of the specimen being tested and the sample preparation for XRD analysis is simple. The disadvantage of XRD as compared to electron diffraction is the small diffraction intensity of X-rays. For that reason, XRD requires a large amount of material and thus the average of the crystallite size is always taken into account (Cullity and Stock, 2001; Hemalatha et al., 2011).

3.2.6 Scanning Probe Microscopy (SPM)

SPM is usually a combination of scanning tunneling microscopy and atomic force microscopy. As compared to other structural characterization methods SPM is a unique characterization method that can analyze three-dimensional (3D) images of a real surface with atomic resolution. This method permits spatially localized measurements of structure and properties of materials. SPM can be used to examine conductive or non-conductive hard or soft solid sample surfaces (e.g. conductors, semiconductors, insulators, magnetic, and opaque materials). The SPM has a wide range of applications such as in nano-indentation, nanolithography, and patterned self-assembly surfaces (Cao, 2004; Bonnell, 2001).

3.2.6.1 Scanning Tunneling Microscopy

The principle of scanning tunneling microscopy is based on electron tunneling. In this microscopy, a sharp tip is positioned above the sample surface (both are either made of a metal or semiconductor) and separated by an insulator or a vacuum. As there is an energy barrier, the electrons cannot move from the tip to the sample surface through the insulator. When a voltage is applied between the two surfaces, the energy barrier shape is changed and the electrons are driven by a force to move across the barrier by tunneling. This will result in the flow of a small current when the distance is sufficiently less. The tip would move above the sample surface in three dimensions accurately and is controlled by an array of piezoelectric. When the tip moves across the sample surface at very small intervals, it scans 3D images of the sample surface with resolution of 0.01 nm in XY direction and 0.002 nm in Z direction. The limitation of scanning tunneling microscopy is that it is applicable to conductive surfaces only because it depends on the tunneling current which is monitored between the tip and sample surface.

3.2.6.2 Atomic Force Microscopy

Atomic force microscopy is a modified form of scanning tunneling microscopy and is applicable to non-conductive surfaces. This microscopy also scans 3D images of the sample surface by moving the tip across the sample surface. However, instead of maintaining a constant tunneling current between the tip and the surface by adjusting the height of the tip as in scanning tunneling microscopy, the atomic force microscopy measures the small upward and downward deflections of the nanoscale tip cantilever while maintaining a constant force of contact.

3.3 DIAGNOSTIC METHODS FOR CHEMICAL CHARACTERIZATION

3.3.1 UltraViolet-Visible (UV-Vis) Spectroscopy

UV-vis spectroscopy is a technique used for identifying qualitatively the presence of colloidal metallic nanoparticles, which shows characteristic absorption spectra. In particular, it is used for measuring the surface plasmon resonance spectra of nanoparticles. In UV-vis spectroscopy, the absorption of light or electromagnetic waves passing through a sample is measured at different wavelengths in the near ultraviolet and visible portions of the spectrum. When sample molecules are exposed to light and having an energy that matches a possible electronic transition within the molecule, some of the light energy will be absorbed as the electron is promoted to a higher energy orbital. The spectrometer records the wavelengths at which absorption occurs, together with the degree of absorption at each wavelength. The resulting spectrum is presented as a plot of absorbance versus wavelength. Absorbance usually ranges from 0 (no absorption) to 2 (99% absorption), and is precisely defined in context with spectrometer operation.

The absorption spectra of the colloidal suspension of nanoparticles could be measured at room temperature using a double beam UV-vis spectrophotometer. Figure 3.1 shows the schematic diagram of a double beam UV-vis spectrometer. This spectrometer works in the wavelength range of 190–1100 nm. The spectrometer consists of a light source, diffraction grating, monochromator with a slit, sample holder (i.e. rectangular type quartz-cuvette),

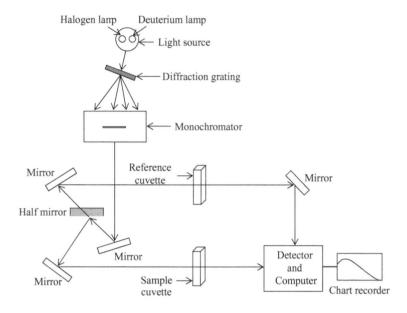

FIGURE 3.1   Schematic diagram of a double beam UV-vis spectrometer.

series of mirrors, and detector. The light source has two lamps— halogen lamp and deuterium lamp—which give an entire visible spectrum plus the near UV so that the wavelength range of about 190–1100 nm can be covered. A diffraction grating splits the light beam at various wavelengths in different directions. The monochromator with a slit separates and allows the light into a very narrow range of wavelengths that will reach the cuvette containing the sample. The cuvette which contains the sample has an internal width of 10 mm (i.e. path length of 10 mm) and a volume capacity of 3.5 ml.

The series of mirrors are used to get the light beam to pass through a reference cuvette and sample cuvette; this allows for more accurate readings. The reference cuvette contains only the solvent (i.e. reference sample) and the sample cuvette contains colloidal suspension samples being studied in the reference sample. Initially, for zeroing the spectrophotometer, i.e. for zero absorbance, the solvent was taken in both the cuvettes and the experimentation could be carried out, and then the test samples could be transferred to the sample cuvette. The detector measures the intensity of the light beam that passes through the sample cuvette, and compares it to the intensity of the light beam that passes through the reference cuvette. Finally, the resulting spectrum is shown as a chart of absorbance of light by the colloidal suspension samples over a specified working range of wavelengths in the visible and the near UV region (Das et al., 2008; Cao, 2004).

3.3.2 Ionic Spectrometry

In ionic spectrometry, high energy beams (order of MeV) of light ions enter into the target sample to a depth of hundreds of nanometers and collide elastically with the atoms of the sample. Thus, the incident ions lose their energies through electronic excitation and ionization of target atoms. Due to the Coulomb repulsion between ions and target atoms, the incident ions are backscattered, which is known as Rutherford backscattering. The energy of backscattered incident ions (E_1) is given by:

$$E_1 = \left(\frac{\sqrt{M^2 - M_0^2 \sin^2 \theta} + M_0 \cos \theta}{M_0 + M} \right) \times 2E_0 \qquad (3.3)$$

By measuring the number and energy of backscattered incident ions, the nature, distribution, and concentration of elements in the sample can be evaluated concurrently (Cao, 2004).

3.4 DIAGNOSTIC METHODS FOR APPLICATION CHARACTERIZATION

3.4.1 Ultrasonic Velocity Measurement

The concentration characterization of colloidal suspension of nanoparticles has been found to be a key point for various applications because nanoparticles have excellent physical, thermal, electrical, catalytic, optical, and antibacterial properties, etc. which play an important role. The concentration of dispersed solid nanoparticles in a liquid media has been measured by various techniques such as Dynamic Light Scattering (DLS), suspended solid meters, UV-vis spectroscopy, Electro Spray Differential Mobility Analysis (ES-DMA), and mass spectrometry. But these techniques work on a very small volume of fluid. Using these techniques, it is also difficult to measure the in-line concentration of a large volume of fluid continuously and therefore they are not acceptable for industrial applications. To meet the shortcomings of the techniques mentioned, the ultrasonic velocity method has been exploited as a new approach (Chakraborty et al., 2011).

The ultrasonic velocity measurement technique can be used to measure the relative concentration of colloidal suspension. The key feature of this method is that it is able to measure concentration of a large amount of fluid both in the laboratory and in-process. Ultrasonic velocity measurement is an established method of finding the adiabatic compressibility of materials, thus this has been an essential part of the science of thermodynamics (Zeemansky, 1957). Because of its flexible and non-destructive character, the ultrasonic technique finds application in several industrial processes such as in level, concentration, and flow measurements of liquids and gases (Povey, 1997; Sheppard, 1994), metal fabrication (Kuttruff, 1991), and controlling the concrete porosity and integrity of metal structures (Povey, 1997).

It is well known that the velocity of sound in a medium is independent of frequency and it depends on the properties of the medium. The Wood equation relates the velocity of sound in homogeneous media to the density and bulk modulus of elasticity (Povey, 1997) as follows:

$$v = \sqrt{\frac{\beta}{\rho}} = \sqrt{\frac{1}{k\rho}} \qquad (3.4)$$

where v is the velocity of sound (m/s), β is bulk modulus (Pa), ρ is density (kg/m^3), and k is adiabatic compressibility (Pa^{-1}) of the media.

However, in the case of mixtures, Wood pointed out that the velocity of sound will be influenced by the mean density and mean compressibility (Wood, 1941). Over the course of time, for colloidal suspensions of solid particles, Urick and Ament (Povey, 1997; Urick and Ament, 1949) have established a model correlating the ultrasonic velocity in suspensions to the mean density and mean compressibility as follows:

$$v_s = \sqrt{\frac{1}{k_{mean}\, \rho_{mean}}} \tag{3.5}$$

where v_s is the ultrasonic velocity in the colloidal suspensions; k_{mean} and ρ_{mean} are the mean compressibility and density of colloidal suspension, respectively, and obtained as follows:

$$k_{mean} = \sum_j \phi_j k_j \tag{3.6}$$

$$\rho_{mean} = \sum_j \phi_j \rho_j \tag{3.7}$$

where ϕ_j is the volume fraction of the jth component in the colloidal suspension; k_j and ρ_j are compressibility and density of the jth component of the colloidal suspension.

In the case of one component dispersed within another, the compressibility and density of the colloidal suspension may be written as:

$$k_{mean} = \phi_1 k_1 + \left(1 - \phi_1\right) k_2, \quad \rho_{mean} = \phi_1 \rho_1 + \left(1 - \phi_1\right) \rho_2 \tag{3.8}$$

Here, the subscripts refer to the constituent components; in particular 1 refers to the dispersed component in a colloidal suspension and 2 to the suspension media. The volume fraction (ϕ_1) of the dispersed component in the colloidal suspension is used to calculate the mass fraction (w) of the same component (Povey, 1997),

$$w = \frac{\phi_1 \rho_1}{\phi_1 \rho_1 + \left(1 - \phi_1\right) \rho_2} \tag{3.9}$$

In Urick and Ament's model, suspensions are assumed as an ideal solution—homogeneous solution—and regarding the propagation of sound in the suspension, the discontinuities are imagined to be non-existent.

Figure 3.2 shows the schematic diagram of ultrasonic velocity measurement setup. The setup consists of an ultrasonic probe (size: 12.7 mm; frequency: 10 MHz; accuracy is within 1 μs), computer controlled pulser/receiver unit (200 MHz frequency), PC data acquisition system software, and motion controller.

When the ultrasonic probe moves to position 1 in the colloidal suspension from the reference point (3 mm below the colloidal suspension), the probe was excited by timed pulses from an ultrasonic pulser-receiver unit and after stimulating the probe a pulse of sound is created. This pulse of sound travels through the colloidal suspensions and it will reflect back from the bottom surface of the beaker. The returning pulse is detected and the resultant time known as acoustic time of flight

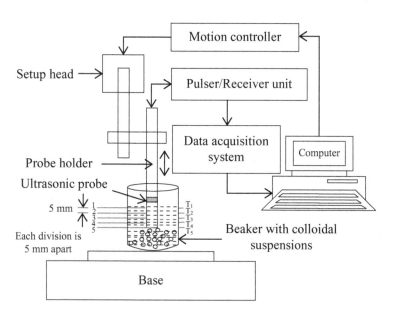

FIGURE 3.2 Schematic of ultrasonic velocity measurement setup.

(T_1) is recorded using the data acquisition system. Similarly, the acoustic time of flight for a different position of the probe (position 2, 3, 4, and 5) will be recorded by moving the probe down consistently by 5 mm from the reference position using the motion controller.

The average acoustic time of flight for colloidal suspension of nanoparticles can be determined as:

$$T_s = \frac{\left(T_1 - T_2\right) + \left(T_2 - T_3\right) + \left(T_3 - T_4\right) + \left(T_4 - T_5\right)}{4} \quad (3.10)$$

The average acoustic time of flight (T_s) can be used to calculate the ultrasonic velocity (v_s) in colloidal suspension according to the Urick's equation (Chakraborty et al., 2011),

$$v_s = \frac{2h}{T_s} \quad (3.11)$$

where h is the length of the fluid column between the successive positions of the probe displaced in the beaker. The volume fraction (ϕ_1) of the nanoparticles can be determined by substituting Equation (3.8) in Equation (3.5) and thus the mass fraction (w) of the colloidal suspension, which is the mass of the nanoparticles dispersed in the suspension medium, can be determined using Equation (3.9).

3.4.2 Thermal and Electrical Conductivity Measurements

The thermal conductivity of colloidal suspension of nanoparticles is measured by the apparatus (schematic shown in Figure 3.3a) built on the principle of the Transient Line Heat-Source (TLHS) method. The TLIIS method involves a long and thin needle probe (size: Ø 2.4 mm and length 100 mm) coated with an electrically insulating layer. The needle is suspended in a colloidal suspension sample (contained in a bottle) symmetrically and an electrical power is applied through it. The increase in temperature of the needle is measured with respect to time. The thermal conductivity (K) of colloidal suspension was measured by measuring the temperature versus time response of the needle to an abrupt electrical impulse and using Fourier's law:

$$K = \frac{q(\Delta \ln t)}{4\pi L(\Delta T)} \tag{3.12}$$

where q is the electrical power, L is the needle length, T is the temperature of the needle at time t. The TLHS method can be used to determine the thermal conductivity of colloidal suspension samples more accurately. The advantages of the TLHS method include simplicity, elimination of natural convection effects, and very fast measurement (Murshed et al., 2008; Nagasaka and Nagashima, 1981).

The electrical conductivity of colloidal suspension of nanoparticles is measured by a 4-ring conductivity electrode meter. The experimental setup of the electrode meter is shown in Figure 3.3b. The setup consists of a 4-ring conductivity probe, bench meter, adjustable handle, stand, beaker containing colloidal suspension sample, and heater. The probe when immersed into the sample to a depth designated by a ring measures the temperature and electrical conductivity of the sample in °C and µS/cm, respectively, at a given instant. The electrical conductivity is determined by Ohm's law. The conductivity of the sample is proportional to its ion concentration because the charge on ions in sample facilitates the conductance of electrical current. However, for measurements above room temperature, the colloidal suspensions sample could be heated up to 90°C by a heater and for measurements below room temperature, the samples could be placed in a freezer chamber (Sarojini et al., 2013; Ganguly et al., 2009).

3.4.3 Viscosity Measurement

The dynamic viscosity of the colloidal suspensions can be measured using a Brookfield viscometer (Model LV DV-II + Pro, USA; working temperature range: −100°C to +300°C, accuracy: ±1%) as shown in Figure 3.4.

This instrument mainly consists of an Ultra Low Adapter (ULA), screen display, control keys, and water bath. The ULA

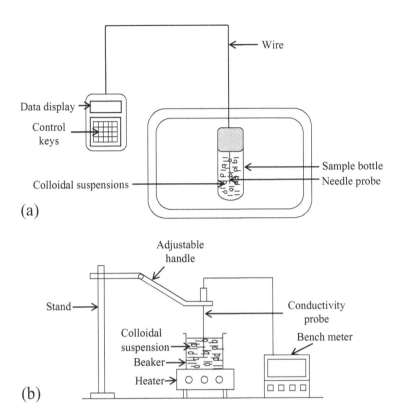

FIGURE 3.3 Experimental setup: (a) Thermal conductivity measurement, (b) Electrical conductivity measurement of colloidal suspensions of nanoparticles.

comprises a cylindrical sample chamber and a spindle which is made of stainless steel attached to the viscometer. The ULA spindle was used at different rpm (i.e. at different shear rates) and can rotate at a maximum speed of 200 rpm inside the chamber containing the sample. The adapter provides accurate viscosity measurements in the range of 1 mPa-s to 2000 mPa-s of a small sample in the order of 2 ml to 16 ml at precise shear rates. The sample chamber fits into the water jacket using a thumb wheel, so that precise temperature control of the sample can be achieved when a circulating water bath is used. The water jacket is connected to

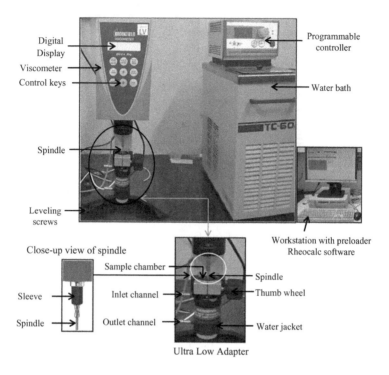

Digital Display
Viscometer
Control keys
Spindle
Leveling screws
Programmable controller
Water bath
Workstation with preloader Rheocalc software

Close-up view of spindle
Sample chamber
Sleeve
Spindle
Inlet channel
Outlet channel
Spindle
Thumb wheel
Water jacket

Ultra Low Adapter

FIGURE 3.4 Brookfield viscometer used for measuring the dynamic viscosity.

the water bath, and the water from the bath enters into the jacket through the inlet channel and exits from the outlet channel. The direct read out of the sample temperature is provided using a Resistance Temperature Detector (RTD) sensor which is embedded in the sample chamber and connected to the viscometer.

The ULA allows the sample chamber to be easily removed and cleaned without disturbing the viscometer setup or water bath. This means that successive measurements can be made under identical conditions. The viscometer was used in conjunction with Brookfield software from Rheocalc. Through Rheocalc software, the viscometer functions are controlled by the computer.

A spindle immersed in the colloidal suspension sample is driven by a torque produced through a spring. The rotation of the spindle generates shear on the sample, causing the sample to flow within the sample chamber. The viscous drag of the sample against the spindle is measured by the spring deflection. Spring deflection is measured with a rotary transducer.

However, to determine the dynamic viscosity of the colloidal suspensions, it is necessary to relate the shear rate and shear stress to the torque and spindle speed. The ULA spindle has two constants, namely, Spindle Shear Rate Constant (SRC) and Spindle Multiplier Constant (SMC) that are given by the manufacturer of the instrument. These constants are used in the calculation of shear rate, shear stress, and viscosity values.

The shear rate ($\dot{\gamma}$) in s^{-1} and shear stress (τ) in N/m^2 can be calculated as:

$$\dot{\gamma} = \mathrm{SRC} \times N \tag{3.13}$$

$$\tau = \frac{T_k \times \mathrm{SMC} \times \mathrm{SRC} \times T_v}{10} \tag{3.14}$$

where
  SRC = Spindle Shear Rate Constant = 1.223
  N = spindle speed, rpm
  T_k = spring torque constant = 0.09373
  SMC = Spindle Multiplier Constant = 0.64
  T_v = viscometer torque (%) expressed as a number between 0
    and 100

The dynamic viscosity of the colloidal sample (μ_d) can be obtained by dividing shear stress by shear rate and is given as:

$$\mu_d = \frac{100}{N} \times T_k \times \mathrm{SMC} \times T_v \tag{3.15}$$

3.5 SUMMARY

In this chapter we discussed the various diagnostic methods used for nanoparticles characterization. Surface characterization methods such as SEM, TEM, and SPM offer the possibility to examine individual nanoparticles. For example, the surface and crystal structure and chemical compositions of nanoparticles can be studied using the surface characterization methods. The XRD method as discussed is used to characterize the collective information of nanoparticles but it does not provide information of individual nanoparticles. The discussion focused on chemical characterization methods of identifying the chemical nature of the colloidal suspensions. Further, diagnostics methods for application characterization were discussed which found applications in the study of nanoparticles.

REFERENCES

Bonnell, D. *Scanning Probe Microscopy and Spectroscopy*, New York: Wiley-VCH, 2001.

Cao, G. *Nanostructures and Nanomaterials: Synthesis, Properties and Applications*, UK: Imperial College Press, 2004.

Chakraborty, S., Saha, S.K., Pandey, J.C. and Das, S. (2011) Experimental characterization of concentration of nanofluid by ultrasonic technique. *Powder Technology*, 210, 304–307.

Cullity, B.D. and Stock, S.R. *Elements of X-Ray Diffraction*, Upper Saddle River, NJ, Prentice Hall, 2001.

Das, S.K., Choi, S.U.S., Yu, W. and Pradeep, T. *Nanofluids: Science and Technology*, USA: Wiley Interscience, 2008.

Ganguly, S., Sikdar, S. and Basu, S. (2009) Experimental investigation of the effective electrical conductivity of aluminium oxide nanofluids. *Powder Technology*, 196, 326–330.

Hemalatha, J., Prabhakaran, T. and Nalini, R.P. (2011) A comparative study on particle-fluid interactions in micro and nanofluids of aluminium oxide. *Microfluid Nanofluid*, 10, 263–270.

Kuttruff, H. *Ultrasonics: Fundamentals and Applications*, The Netherlands: Elsevier Science, 1991.

Murshed, S.M.S., Leong, K.C. and Yang, C. (2008) Thermophysical and electrokinetic properties of nanofluids – a critical review. *Applied Thermal Engineering*, 28, 2109–2125.

Nagasaka, Y. and Nagashima, A. (1981) Absolute measurement of the thermal conductivity of electrically conducting liquids by the transient hotwire method. *Journal of Physics E: Scientific Instruments*, 14, 1435–1440.

Povey, J.W.M. *Ultrasonic Techniques for Fluids Characterization*, USA: Academic Press, 1997.

Sarojini, K.G.K., Manoj, S.V., Singh, P.K., Pradeep, T. and Das, S.K. (2013) Electrical conductivity of ceramic and metallic nanofluids. *Colloids and Surfaces A: Physicochemical and Engineering Aspects*, 417, 39–46.

Sheppard, T.J. (1994) Solid state gas metering – the future. *Flow Measurement and Instrumentation*, 5, 103–106.

Urick, R.J. and Ament, W.S. (1949) The propagation of sound in composite media. *Journal of Acoustical Society of America*, 21, 115–119.

Wang, Z.L. *Reflected Electron Microscopy and Spectroscopy for Surface Analysis*, Cambridge, Cambridge University Press, 1996.

Wood, A.B. *A Textbook of Sound*, UK: Bell & Sons, 1941.

Yao, N. and Wang, Z.L. *Handbook of Microscopy for Nanotechnology*, Boston, MA: Kluwer Academic Publishers, 2005.

Zeemansky, M.W. *Heat and Thermodynamics*, USA: McGraw-Hill, 1957.

CHAPTER 4

A Novel Approach for Nanoparticles Synthesis— EDMM System

4.1 INTRODUCTION

A wide variety of methods have been used for the synthesis of nanoparticles in the form of or without colloids through mechanical, liquid phase reaction, and vapor phase reaction routes. However, the synthesis of nanoparticles using a corona discharge micromachining method—Electrical Discharge Micromachining (EDMM)—has not been explored. Therefore, this chapter is first focused on a novel approach for synthesis of nanoparticles using EDMM. EDMM is a single-step vapor phase reaction method in which nanoparticles are synthesized in the base fluids directly and no redispersion process is required, and hence reduces the production cost. This method can be used to synthesize nanoparticles of any conducting materials concentrating on the size, distribution, aggregation, cost, and high yield compared to other methods. Prior to the realization of EDMM and its approach for synthesis of nanoparticles, it is necessary to understand the non-conventional machining processes in brief because EDMM is one such process. The mechanisms of material removal rate involved in non-conventional machining processes and superiority of these processes over conventional machining processes are discussed. In addition, the importance of conceptual realization of basic electrical discharge machining is detailed in this chapter. Next, in this chapter, the discussion is focused on the development and fabrication of an indigenous prototype EDMM system coupled with an ultrasonicator and piezoactuated tool feed system for the purpose of synthesis of nanoparticles. The advantages of the developed EDMM system as compared to the commercially available EDMM systems would include less cost, it provides key functionalities that are needed for research purposes, it allows free selection of the operative parameters such as duty cycle, frequency of electrical pulses, and voltage in the required range, and it is able to generate single pulse discharges.

4.2 NON-CONVENTIONAL MACHINING PROCESSES

In today's modern world, customer requirements on the products/processes are stringent, such as extraordinary properties of the materials, complex three-dimensional geometries, miniature features, nano level surface finish, multifunctional properties with ease of operations, and low cost. Competition exists worldwide to meet customer requirements efficiently and effectively in the stipulated period of time. Currently many new engineering materials have been developed to meet customer requirements with high standards. Some of the newer engineering materials are nitroalloy, hastalloy, nimonics, carbides, heat resistant steel, waspalloy, etc. These materials find wide applications in aerospace, nuclear engineering, and other industries owing to high strength to weight ratio, high hardness, and high heat resisting quality.

However, machining of these materials is difficult and even impossible with conventional machining processes. Even if it is possible with the advanced machine tools such as Numerical Control/Computer Numerical Control/Direct Numerical Control/ Machine Centers etc., these processes involve thousands of slide movements, high material removal rate, and high speed of operations to machine 3D geometries with required accuracy and precision. The alternative is to develop advanced machining processes known as non-conventional (non-traditional) machining processes, which are used to produce any intricate profiles on various engineering materials—metals and alloys with high accuracy without any subsequent processing steps. So, these processes eliminate further processes like finishing steps as seen in conventional machining processes, thereby reducing the time duration in production and it is not possible to easily meet high aspect ratios of 100:1 in conventional processes.

The non-conventional machining processes are those processes which remove the material using some form of energy namely mechanical, thermal, chemical, and electrochemical with no constraint on the size of the component being machined. These processes are diverse in nature and differ from each other by their characteristic features, operations, and fields of application. Table 4.1 shows the classification of non-conventional machining processes. The listed classification is based on different types of energy and mechanisms used for material removal (Ghosh and Mallick, 2010; Koc and Ozel, 2011; Takahata, 2009; Payal and Sethi, 2003; Jain, 2010). The non-conventional machining processes have unlimited capabilities over conventional processes in terms of machining harder materials, compactness, reducing cost of machining etc. Most of these processes are controlled automatically thereby giving simplicity, reliability, and repeatability resulting in wider acceptance of the processes. Many of these processes are automated with vision systems, laser gauges, and other inspection systems. Also they have the capability to adjust the various process parameters in order to get better output results.

TABLE 4.1 Classification of Non-Conventional Machining Processes

Energy type	Mechanisms of material removal	Energy source	Process	Common applications
Mechanical	Mechanical erosion of workpiece material using abrasive particles	Mechanical motion	USM	Round and irregular holes, impressions, etc.
		Pneumatic	AJM	Drilling, cutting, deburring, etc.
		Hydraulic	WJM	Paint removal, cleaning, cutting frozen meat, etc.
		Hydraulic	AWJM	Peening, cutting, textile, leather industry, etc.
Thermal	Melting and evaporation of workpiece material	Electric spark	EDM	Holes in nozzles and catheters, channel cutting, curved surfaces, etc.
		High speed electrons	EBM	Drilling fine holes, cutting contours in sheets, etc.
		Powerful radiation	LBM	Drilling fine holes, cladding, etc.
		Ionized substance	IBM	3D patterning in IC circuits, micro tools and micro dies fabrication, etc.
		Ionized substance	PAM	Cutting plates, etc.
Chemical	Corrosive reaction	Corrosive agent	CHM	Pockets, contours, MEMS devices, etc.
Electro-chemical	Ion displacement	Electric current	ECM	Blind holes, cavities, etc.

(Continued)

TABLE 4.1 (CONTINUED)   Classification of Non-Conventional Machining Processes

Energy type	Mechanisms of material removal	Energy source	Process	Common applications
Hybrid machining processes				
Electro-chemical and thermal	Melting and evaporation + chemical etching	Electrical discharges and chemical	ECSM	Holes, grooves, channels, complex shape contours, etc.
Chemical and mechanical	Chemical reaction + mechanical abrasion	Abrasive slurry with chemicals	CMP	Polishing of Au and Ti, optoelectronic components, etc.
Electro-chemical and mechanical	Electrolytic reaction + grinding action	Electric current and mechanical motion	ELID	Small hole grinding, fine finish of hard and brittle materials, etc.

USM—Ultrasonic Machining; IBM—Ion Beam Machining; AJM—Abrasive Jet Machining; PAM—Plasma Arc Machining; WJM—Water Jet Machining; CHM—Chemical Machining; AWJM—Abrasive Water Jet Machining; ECM—Electro Chemical Machining; EDM—Electrical Discharge Machining; ECSM—Electro Chemical Spark Machining; EBM—Electron Beam Machining; CMP—Chemo-Mechanical Polishing; LBM—Laser Beam Machining; ELID—Electrolytic In-Process Dressing.

Due to reasons mentioned, the demand for non-conventional machining processes is gaining wider acceptance among manufacturing engineers, product designers, and metallurgical engineers (Benedict, 1987).

However, among the non-conventional machining processes, Electrical Discharge Machining is one of the thermal energy based non-conventional processes used to machine various exotic conducting engineering materials such as nitroalloy, hastalloy, nimonics, carbides, ceramics, etc. along with the general purpose engineering materials such as copper, aluminium, stainless steel, etc. to very high accuracy and precision. This method has become significant for complexity of workpiece shape, size, and surface integrity. The conceptual realization of basic Electrical Discharge Machining is detailed in the following section.

4.3 ELECTRICAL DISCHARGE MACHINING (EDM)

EDM removes the material from the workpiece electrode (anode) and the tool electrode (cathode) by rapidly recurring spark discharges (corona discharges); both the electrodes are immersed in a dielectric fluid. The sparks, occurring at high frequency, continuously and effectively remove the material from both the electrodes in the form of debris by melting and evaporation. This process has become indispensable in modern manufacturing industries because of its ability to obtain adequate material removal rate, produce complex features with high degree of accuracy, and minimize damage to the material properties.

4.3.1 Salient Features of EDM

The salient features of EDM are:

- There is no physical contact between the tool and the workpiece. Hence the workpiece is not subjected to contact stresses.

- Hardness and fragileness of the materials are no barrier to machining.

- Physical and metallurgical properties of the material have no barrier.

- Generation of three-dimensional complex features and implications of size have no effect.

- Capable of machining difficult-to-machine conductive materials such as ceramics, carbides, super alloys, composites, heat resistant steels, etc. irrespective of their mechanical properties.

- The process can easily be automated.

4.3.2 EDM Cell

The schematic diagram of the EDM cell is shown in Figure 4.1a. It consists of a tool electrode (cathode) and workpiece electrode (anode) connected to a DC power supply through a Resistor-Capacitor (RC) circuit in general. The other types of pulse generation circuits that can be used in EDM are rotary impulse, vacuum tube, transistor, and single pole dual-throw switch types. These pulse generation circuits are available in various literatures (Kunieda et al, 2005; Ghosh and Mallick, 2010).

The tool electrode and workpiece electrode are separated by a small gap known as a spark gap and are submerged in a dielectric fluid. The most commonly used tool materials are of a conducting type such as copper, aluminium, brass, graphite, silver, tungsten carbides, copper-tungsten alloys, silver-tungsten alloys, etc. The workpiece materials are also conducting in nature such as copper, aluminium, silver, graphite, steel, super alloys, semiconductors, ceramics, etc. The most commonly used dielectric fluids are De-Ionized (DI) water or hydrocarbon fluids such as kerosene, paraffin oil, and transformer oil. DI water was found to be more favorable other than hydrocarbon fluids. This is because it gives a higher material removal rate which in turn can produce a large amount of debris (Nguyen et al., 2012).

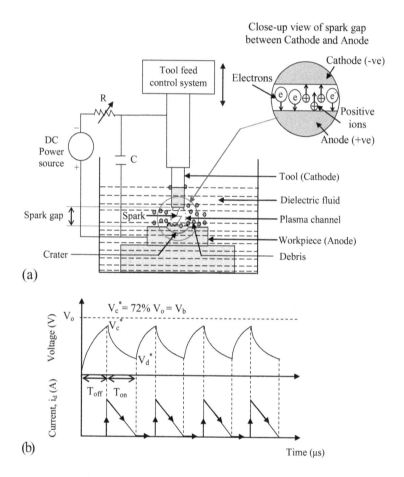

FIGURE 4.1 (a) Schematic diagram of EDM cell, (b) Waveform of voltage and current characteristics.

4.3.3 Mechanism of EDM

The basic mechanism of material removal in EDM is as follows:

At the beginning, an open circuit voltage (V_o) is supplied between the tool and the workpiece. The capacitor is charged through the resistor from the supplied open circuit voltage and would continue to charge until it reaches the breakdown voltage of the dielectric fluid. The charging time of the capacitor is known as pulse interval/pulse off time (T_{off}), as shown in the waveform of voltage and current characteristics of Figure 4.1b. When the voltage builds up to some predetermined value, electrons break loose from the tool (cathode) and are accelerated towards the workpiece (anode). During their travel within the spark gap, the electrons collide with the neutral molecules of dielectric fluid causing ionization which results in the formation of a plasma channel between the two electrodes.

Since the electrical resistance of the plasma channel is much less, all of a sudden a large number of electrons will move from the tool to the workpiece and positive ions from the workpiece to the tool. This is known as an avalanche of electrons. At this point, once the voltage across the capacitor (V_c^*) has reached the breakdown voltage (V_b), i.e. when $V_c^* = 72\%$ $V_o = V_b$, the current (i_d) starts to flow across the electrodes. When the current starts to flow, the voltage drops to the discharge voltage (V_d^*) and the current (i_d) rises to its peak, as shown in Figure 4.1b. Such a phenomenon can be normally seen as corona discharge (sudden discharge of electrons in the form of light/spark discharge). However, the spark discharge can take place up to the duration of T_{on}, known as discharging time of the capacitor/pulse on time (Figure 4.1a). This indicates that the current continues to flow until almost all the energy is drawn out of the capacitor, at which time the current stops and again charging of the capacitor continues. As a result of the spark, a very high temperature of the order of 10,000°C to 12,000°C is produced causing melting and vaporization of the work material at the point of discharge leaving behind a tiny crater on the work surface. A small amount of the vaporized material in the form of debris is dispersed into the space surrounding the electrodes.

As soon as this happens, the gap between the electrodes changes and therefore, the tool feed control system is used to feed the tool in order to maintain uniform spark gap. The commonly used tool feed control systems in EDM are servo motor, stepper motor, solenoid control, electromagnetic motor, inchworm actuator, electromagnetic actuator, macro/micro dual feed drive, and piezoactuator. The major operating parameters of the EDM process are open circuit voltage, peak current, pulse on time (pulse duration), pulse off time (pulse interval), duty cycle, and pulse frequency. A detailed explanation of these parameters is available in text books (Davim, 2013; Koc and Ozel, 2011; Ghosh and Mallick, 2010).

4.4 ELECTRICAL DISCHARGE MICROMACHINING (EDMM)

It has been known that the EDM process is capable of machining complex blind cavities, deep holes, and extremely complicated shapes on various engineering materials. As such, EDM has found widespread applications in many industrial domains, such as mould and die manufacturing, small and burr-free hole drilling in turbine blades, making spool valves and key ways for cylinder, etc. However, with growing trends towards miniaturization of machined parts, developments in the area of Micro-Electro-Mechanical-Systems (MEMS), and requirements for micro-features in difficult-to-machine engineering materials, micromachining with EDM (i.e. EDMM) has become increasingly important.

EDMM is the process of removing material from the electrically conductive components irrespective of their size in the form of debris usually of size in the range of micron (1 μm $\leq$ dimension $\leq$ 999 μm) by using precisely controlled sparks between the electrodes in the presence of a dielectric fluid. It is the most promising and cost-effective advanced micromachining method because of its noncontact nature of machining capability due to micro-sized tools. This process involves very low input energy in the order of 10^{-7} Joule resulting in a lower unit removal capability. The process is capable of machining intricate micro-features on conducting and semiconducting materials in the form of debris at a lower cost and at less time in micron/nano level with the required accuracy and precision.

EDMM has found promising applications in various industries, such as machining of micro holes in fuel injection automotive nozzles, filters for food processing and textile industries, cooling holes in aircraft turbine blades, spinneret holes for synthetic fibers, fiber-optics, micro opto-electronic components, micro-mechatronic actuator parts, catheters, needles, and other medical device components. It can be also used for machining of micro channels in reactors, MEMS, and fluidic devices, etc. Consequently, the use of EDMM for micro-scale manufacturing became an inevitable and popular micromachining process.

The basic mechanism of the EDMM process is similar to that of the EDM process. Even though the basic mechanism is the same, there are significant differences between EDM and EDMM in the size of the tool, pulse generator, spark gap, specific energy of material removal, and the operating parameters. Table 4.2 shows the differences between EDM and EDMM based on the available literature (Davim, 2013; Dhanik and Joshi, 2005; Wong et al., 2003; Uhlmann et al., 2005; Gurule and Pansare, 2013).

TABLE 4.2   Differences between EDM and EDMM

Parameters	EDM	EDMM
Size of the tool	Greater than 1 mm	Less than 1 mm
Pulse generator	RC relaxation type	RC single pulse discharge
Current and gap voltage	Constant throughout the discharge	Change during the discharge
Spark gap	Greater than 10 μm	Less than 10 μm
Specific energy of material removal	High (of the order of $\sim 10^{-5}$ J)	Low (of the order of $\sim 10^{-7}$ J)
Efficiency of process	Low	High
Crater size per single spark discharge	Greater than 5 μm	Less than 2.5 μm
Precision and accuracy of final products	Greater than 5 μm	Less than 5 μm
Open circuit voltage	$\sim$ 40 to 400 V	$\sim$ 10 to 120 V
Peak current	Greater than 3 A	Less than 3 A
Pulse on time	$\sim$ 0.5 to 8 ms	$\sim$ 50 ns to 100 μs

4.4.1 EDMM Approach for Nanoparticles Synthesis

As discussed earlier, it was found that the EDMM method has been used in the machining of miniature features/parts by very small unit material removal phenomenon. The unit removal of material from the electrodes is in the form of debris at micro/nano level. The debris generated during micromachining is being treated as unwanted particles and they are flushed away from the spark gap by dielectric fluid.

So far, most of the research has concentrated on the production of miniature components by the EDMM system but attention has not been focused on the debris and its applications as nanoparticles in the field of nanotechnology. Therefore, in this book attention has been focused on the synthesis of debris in the dielectric fluid using the indigenous EDMM system. The collection, characterization, and application suitability of these colloidal suspensions of debris (colloidal suspensions of particles) have gained significant societal attention. In this chapter, we are concerned with the development and fabrication of an indigenous prototype EDMM system coupled with an ultrasonicator and piezoactuated tool feed system for the purpose of synthesis of nanoparticles. The indigenous EDMM system could be simple, compact, cost-effective, and versatile. The realization of various units of the developed EDMM system is discussed in the following section.

4.5 PROTOTYPE EDMM SYSTEM

The block diagram and the photographic view of the indigenous prototype EDMM system is shown in Figures 4.2a and b, respectively. The system consists of a tool electrode (cathode) and a workpiece electrode (anode) separated by a small gap known as a spark gap, surrounded by dielectric fluid, and is connected to the DC power supply (0–32 V). In the system, a tool feed controller unit is used for tool feed control through a piezoactuator to maintain a constant spark gap between the electrodes. This is achieved based on the average gap voltage as a feedback signal measured using a gap voltage sensor. Pulses of specified frequency in the

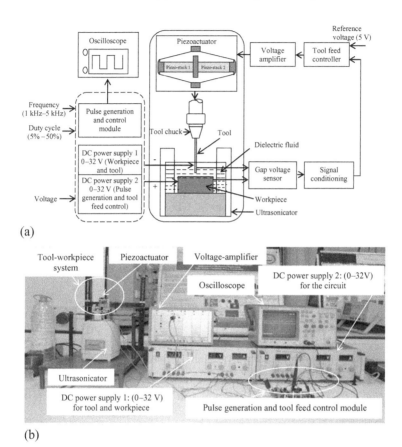

(a)

(b)

FIGURE 4.2   Indigenous prototype EDMM system: (a) Block diagram, (b) Photographic view.

range of 1 kHz to 5 kHz and duty cycle in the range of 5% to 50% are applied to the electrodes using the developed pulse generation and control circuit. An oscilloscope is used to monitor and capture the pulse discharge waveform during machining.

In the present EDMM system, De-Ionized (DI) water is used as the dielectric fluid. This is because, in most of the literature, DI water was used as a common base fluid for formulating colloidal suspension of nanoparticles. Table 4.3 shows the detailed features of the indigenous prototype EDMM system.

TABLE 4.3   Features of the Indigenous Prototype EDMM System

Features	Specification
Workpiece	Electrically conducting materials such as Cu, Al, Ag, Au, W, etc. or semiconductors such as silicon, graphite, etc. Dimension: maximum of 40 mm × 40 mm × 0.5 mm
Tool	Electrically conducting materials such as Cu, Al, Ag, Au, W, etc. or semiconductors such as silicon, graphite, etc. Dimension: diameter up to 900 μm, length up to 50000 μm
Dielectric fluid	De-Ionized (DI) water
Piezoactuator for tool feed control	Type of control: analog Manufacturer: Cedrat Technologies, France Maximum displacement: 400 μm Nominal load carrying capacity: 3.8 kg Resonance frequency: 495 Hz Response time: 1.01 ms Drive system: 0–150 V
Pulse generation and control circuit	Transistor type Pulse frequency range: 1–5 kHz Duty cycle range: 5%–50%
Power supply to the tool and workpiece	Voltage: 0–120 V Peak current: less than 3 A
Isolated power supply to the circuit	Voltage: 12–15 V Peak current: less than 0.5 A

The various units involved in the prototype EDMM system are briefly discussed in the following subsections.

4.5.1 Ultrasonicator

The ultrasonicator has a frequency generator and Lead Zirconate Titanate (PZT) transducer that produce ultrasonic vibrations with frequency 55 kHz during the experiment to achieve uniform dispersion and to avoid the aggregation of particles in the colloidal suspensions. An ultrasonicator can also aid the removal of debris from the spark gap region efficiently and avoid the accumulation of debris in the spark gap. This results in stable spark discharges and increased machining efficiency.

4.5.2 Piezoactuator

A piezoactuator is a device that can produce mechanical energy when an electric field is applied and vice-versa. It is a durable and inert component, known for its quick reaction speed and reproducibility of the travelled distance. The piezoactuator contains piezo-elements which are made of Lead Zirconate Titanate (PZT) material and each element can expand or contract by about 0.1% to 0.2% of their length in one direction. Thus, it can be used in applications in which very small but precise displacements are needed, having a short reaction time such as the μ-EDM process.

The piezoactuator in the EDMM system is used for tool feed to maintain a constant spark gap between the electrodes. It avoids the sparks occurring with longer delay time and the short circuiting of electrodes. It exhibits fast response characteristics (response time is less than a millisecond), high stiffness, and low wear and tear. Table 4.4 shows the main characteristics of the piezoactuator. It consists of two piezostacks arranged in series. Each piezostack consists of 196 piezo wafers of 0.1 mm thickness and 5×5 mm^2 area. These two piezostacks are housed inside a flexural link to amplify the displacement of the actuator. The upward and downward movement of the piezoactuator is caused by the flexural link. The maximum vertical motion of the tool is 400 μm at 150 V.

TABLE 4.4   Characteristics of the Piezoactuator

Characteristics	Values
Mass (gm)	19.0
Dimension (mm)	$48.4 \times 11.5 \times 13.0$
Stiffness (N/m)	0.1×10^6
Voltage range (V)	0–150
Capacitance (F)	4.0×10^{-6}
Resolution (nm)	4.0
Number of piezostacks	2
Number of wafers (each piezostack)	196
Thickness of each wafer (mm)	0.1
Area of each wafer (mm^2)	25

To get 150 V from an available DC Power source (0–32 V) and voltage amplifier (20×), a maximum voltage drawn from the DC power supply is about 7.5 V (7.5 × 20 = 150 V). If the amplified signal is above 150 V, a protective diode and the power amplifier saturation will clamp the signal. Hence the voltage output from the amplifier to the actuator is always within the safe operating range of the actuator (0–150 V).

4.5.3 Pulse Generation and Control Module

The Metal-Oxide Semiconductor Field Effect Transistor (MOSFET) based pulse width modulated-type pulse generation and control module is used to provide rectangular pulses at a particular frequency and duty cycle to the electrodes' tool and workpiece. This module is driven by a DC power supply 2. The advantage of using the power supply 2 is that the open circuit voltage which is applied between the electrodes using DC power supply 1 need not be attenuated. The module can handle large current and thus, the material removal rate will be much higher. The pulse generation and control circuit consist of a Pulse Width Modulator (PWM), which produces rectangular pulses at different frequencies in the range of 1 kHz to 5 kHz and duty cycles 5% to 50%. The duty cycle and frequency of the pulse discharge are controlled by adjusting the values of resistance and capacitance using the potentiometers which are connected to the Pulse Width Modulator (PWM). The PWM adjusts the pulse duration in a pulse train relative to a control signal. The main advantage of the PWM is that power loss in the switching devices is very low. In this module, arrangement is made to change the pulse duration by varying the duty cycle and frequency at any point of time during the EDMM process depending on the machining conditions required.

4.5.4 Tool Feed Control Module

The tool feed control module is used to control the tool feed through a piezoactuator based on the average gap voltage as the feedback signal. As a result, a constant spark gap of about 10 μm between the electrodes can be maintained to sustain spark discharges for an efficient EDMM operation. The module comprises a voltage comparator, a relay single pole dual-throw switch, and a ramp generator which uses an operational amplifier. In the EDMM system, the tool is moved towards the workpiece using an adjustable knob till the gap between the tool and the workpiece reaches a value equivalent to the spark gap (10 μm). When the required spark gap is reached, sparks are produced at the supplied pulse frequency, resulting in removal of material from the electrodes in the form of debris by melting and evaporation. This results in an increase in the spark gap and also the average gap voltage. The gap voltage signal is filtered and is compared with the reference voltage (+5 V) by a voltage comparator corresponding to the set spark gap (10 μm). The comparator output is connected to a relay which switches high and low signal to the ramp generator. The ramp generator produces voltage signal with either positive or negative slope from the current voltage level by using an operational amplifier. This signal is amplified through a voltage amplifier connected at the output of the ramp generator and supplied to the actuator which in turn feeds the tool and controls the spark gap.

4.6 SUMMARY

In this chapter, the importance of non-conventional machining processes over conventional machining processes was discussed. The use of corona discharge micromachining—EDMM—as an inevitable micro-scale manufacturing method by surpassing EDM was briefly explained. The differences between EDM and EDMM in terms of various parameters are highlighted. The discussion focused on the approach for synthesis of nanoparticles using EDMM. Further, in this chapter, the development of an indigenous prototype EDMM system coupled with an ultrasonicator and piezoactuated tool feed system for the purpose of synthesis of nanoparticles was discussed in detail. The ultrasonicator can achieve uniform dispersion and avoid the aggregation of particles generated during the experiment. It can also help in removing the debris from the spark gap region efficiently and avoids the accumulation of debris in the spark gap during the EDMM process resulting in increased machining efficiency. The piezoactuator is used for tool feed control to maintain uniform spark gap during the process. The gap voltage based feedback signal was found to be suitable for tool feed control through the piezoactuator in the EDMM system.

REFERENCES

Benedict, G.F. *Nontraditional Manufacturing Processes,* USA: CRC Press, 1987.

Davim, J.P. *Nontraditional Machining Processes: Research Advances,* 1st edition, Chapter 4, UK: Springer-Verlag, 2013.

Dhanik, S. and Joshi, S.S. (2005) Modeling of a single resistance capacitance pulse discharge in micro-electro discharge machining. *Journal of Manufacturing Science and Engineering,* 127, 759–767.

Ghosh, A. and Mallick, A.K. *Manufacturing Science,* 2nd edition, East-West Press (India) Pvt. Ltd, 2010.

Gurule, N.B. and Pansare, S.A. (2013) Potentials of micro-EDM. *IOSR Journal of Mechanical and Civil Engineering,* 6, 50–55.

Jain, V.K. *Introduction to Micromachining,* India: Narosa Publishing House Pvt. Ltd, 2010.

Koc, M. and Ozel, T. *Micro-Manufacturing: Design and Manufacturing of Micro-Products,* 1st edition, Chapter 10, USA: John Wiley & Sons, 2011.

Kunieda, M., Lauwers, B., Rajurkar, K.P. and Schumacher, B.M. (2005) Advancing EDM through fundamental insight into the process. *CIRP Annals - Manufacturing Technology,* 54 (2), 64–87.

Nguyen, M.D., Rahman, M. and Wong, Y.S. (2012) An experimental study on micro-EDM in low-resistivity deionized water using short voltage pulses. *International Journal of Advanced Manufacturing Technology,* 58, 533–544.

Payal, H.S. and Sethi, B.L. (2003) Non-conventional machining processes as viable alternatives for production with specific reference to electrical discharge machining. *Journal of Scientific and Industrial Research,* 62, 678–682.

Takahata, K. *Micro Electronic and Mechanical Systems,* 1st edition, Chapter 10, Croatia: INTECH, 2009.

Uhlmann, E., Piltz, S. and Doll, U. (2005) Machining of micro/miniature dies and moulds by electrical discharge machining-recent development. *Journal of Materials Processing Technology*, 167, 488–493.

Wong, Y.S., Rahman, M., Lim, H.S., Han, H. and Ravi, N. (2003) Investigation of micro-EDM material removal characteristics using single RC-pulse discharges. *Journal of Materials Processing Technology*, 140, 303–307.

CHAPTER 5

Synthesis, Characterization, and Application Suitability

5.1 INTRODUCTION

The indigenously developed Electrical Discharge Micromachining (EDMM) prototype can be used to produce any conducting nanoparticles in general, and in particular it can be used to synthesize copper and aluminium nanoparticles. Currently, copper and aluminium nanoparticles have been gaining remarkable research interest and market attention as was seen in Chapter 1, because they are inexpensive, exhibit unique physical, thermal, catalytic, electrical, antibacterial, and optical properties, and are widely used in various potential applications. Indicatively, such applications include lubricants, cooling systems, Micro-Electro-Mechanical-Systems (MEMS) devices, biomedical, catalysis, rocket propellants, explosives, and so on. The EDMM is a single-step method to synthesize colloidal suspensions i.e. nanoparticles in base fluid while machining the conducting materials. The indigenously developed prototype EDMM system is much more economical compared to the commercially available Electrical Discharge Machining (EDM) system, and the parameters of interest are also easily varied and controlled. Pulse generation and control module, and the tool feed control module incorporated in the system play a major role during the process.

In this chapter, the discussion is focused on the experimental parameters and procedure adopted for synthesis of copper and aluminium nanoparticles using the developed prototype EDMM system. The characterization results of the synthesized colloidal copper and aluminium nanoparticles along with a detailed discussion are presented. The possible methods to overcome agglomeration of particles and achieve stable dispersion are also discussed. Further, the development of a thermal management setup to study the heat transfer application of the synthesized colloidal nanoparticles is demonstrated in the chapter.

5.2 SYNTHESIS OF COPPER NANOPARTICLES

5.2.1 Experimental Parameters

The experimental process parameters such as open circuit voltage, peak current, and pulse duration are typically used during the machining process. In the EDMM system, the open circuit voltage can be varied from 0 to 30 V and the peak current is kept below 3 A. The pulse duration can be set by changing the frequency and the duty cycle. Table 5.1 shows the experimental parameters used for the synthesis of copper nanoparticles.

5.2.2 Experimental Procedure

In the present study, 200 ml of DI water was taken in an ultrasonicator. The required open circuit voltage of 30 V is set for the experiment using a DC power supply. The peak current is limited to 2 A to prevent the power supply from short circuiting. The feedback voltage should lie between 0 and 7.5 V to prevent any damage to

TABLE 5.1 Experimental Parameters Used for the Synthesis
of Copper Nanoparticles

Parameters	Values
Open circuit voltage	30 V
Peak current	2 A
Frequency	5 kHz
Duty cycle	30%
Pulse duration	60 μs
Workpiece electrode (anode)	Copper plate of 300 μm thickness (99.0% purity)
Tool electrode (cathode)	Copper wire of 900 μm diameter (99.0% purity)
Dielectric fluid (base fluid)	De-Ionized (DI) water—200 ml
Environment	Room temperature

the voltage amplifier. From the initial position the tool electrode is moved towards the workpiece electrode using an adjustable knob until the gap between the electrodes reaches a value equivalent to the spark gap. The MOSFETs switch the high voltage applied between the tool electrode and workpiece electrode at the specified frequency and duty cycle. Due to very high potential gradient across the electrodes, a plasma channel is formed and a spark discharge takes place. Due to the spark, localized intense heat is produced which increases workpiece temperature in a narrow zone, causing melting and vaporization of the work material. The vaporized material condenses and results in the formation of debris. As the workpiece is eroded, and the spark gap subsequently increased, the tool is fed automatically by the piezoactuator so that the machining at each point would continue without any interruption.

In this study, machining is done at various points on the workpiece by realigning the tool. The debris of copper thus produced is suspended in the dielectric fluid and the obtained colloidal suspension of copper particles was collected in glass vials. Characterization of the generated colloidal suspension of copper particles was carried out using various diagnostic studies and the results are discussed in the following sections.

5.3 STRUCTURAL CHARACTERIZATION OF SYNTHESIZED NANOPARTICLES

5.3.1 Colloidal Copper Nanoparticles

The colloidal suspension of synthesized copper particles was characterized to recognize size, shape and distribution, chemical composition, crystal structure, concentration, thermal conductivity, and viscosity. The size, shape, and distribution of the colloidal copper nanoparticles were examined through a Transmission Electron Microscope (TEM) three hours after production. The composition of the colloidal copper nanoparticles is analyzed using EDAX. The crystal structure of the copper nanoparticles synthesized in the dielectric fluid is studied by Selected Area Electron Diffraction (SAED) pattern. This makes it possible to confirm the element of nanoparticles synthesized.

For TEM analysis, the synthesized colloidal particles sample is prepared on the copper grid with proper care. The colloidal particles in a glass tube collected after synthesis using the EDMM system were sonicated for about 1 hr, to ensure uniform dispersion of the nanoparticles in the sample. Immediately 1–2 drops of the colloidal suspensions are placed onto a 50-mesh carbon-coated copper grid of size 3.05 mm kept on a micro-filter paper. Then the sample was kept drying at an ambient temperature for about 1 hr. It forms a thin layer of precipitate over the copper grid; the size and morphology of the particles retained over the grid was visualized using the TEM instrument.

5.3.1.1 Size, Shape, and Distribution in Pure DI Water

The TEM images of copper nanoparticles synthesized in pure DI water is shown in Figure 5.1a. The images illustrate that the copper nanoparticles are almost spherical in shape. The particles size lies between 600 nm and 1100 nm. The mean size of the particle was evaluated to be 880 nm. The TEM images showed the presence of large sized copper particles with defined singles. Such large sized synthesized particles are attributed to agglomeration of copper nanoparticles resulting from the combination of the

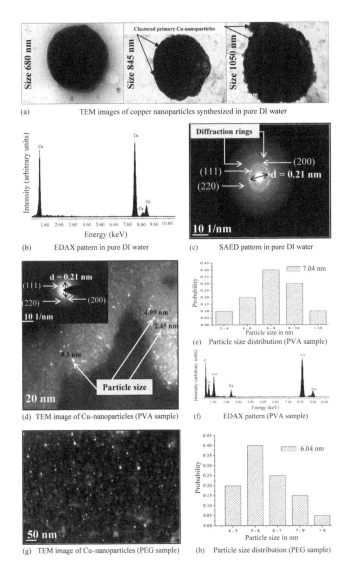

FIGURE 5.1 Size, shape, and distribution of colloidal copper nanoparticles: (a) TEM images of copper nanoparticles synthesized in pure DI water, (b) EDAX pattern in pure DI water, (c) SAED pattern in pure DI water, (d) TEM image of Cu nanoparticles (PVA sample), (e) Particle size distribution (PVA sample), (f) EDAX pattern (PVA sample), (g) TEM, (h) Particle.

van der Waals attraction force and Brownian motion during condensation and nucleation of the metallic copper vapors.

The sub-micron particle of size 680 nm shows a dense structure throughout. This may be either attributed to the collision of nucleated nanoparticles with each other or agglomeration of particles that results in a larger size. However, the particles of size 845 nm and 1050 nm do not show much dense structure at the periphery indicating the clustering of primary copper nanoparticles. In the Cu-DI water colloidal suspension sample, the particles size distribution histogram is difficult to show because of the presence of a lower number of large sized particles in the sample used for the TEM analysis.

The EDAX pattern of copper nanoparticles in pure DI water is shown in Figure 5.1b. The pattern clearly shows the distinct elemental peaks of copper particle because of the transition of electrons to the L and K shells in copper atoms. This indicates the presence of copper particles synthesized in the pure DI water. As copper contains high atomic density crystallographic facets, four characteristic peaks of copper particle were found to have appeared at different crystal facets. Moreover, it is demonstrated that oxides of copper are not observed. In addition, no additional impurities were detected in the EDAX pattern. Thus, the EDAX analysis of the Cu-DI water sample indicates that the sample contains only metallic copper, with no oxide layer/particles. Figure 5.1c shows the SAED pattern of the copper nanoparticle in pure DI water. The figure shows a sequence of spots; every spot is related to a pacified diffraction condition of the particles' crystal structure. The pattern shows clear diffraction rings; the distance from the centre to each concentric ring was measured. The first ring diameter from the centre yield lattice (d) spacing of 0.21 nm perfectly matched with the FCC crystalline copper lattice distance of 0.208 nm for the (111) plane. Similarly, the d spacing for the next two rings were found to be 0.18 nm and 0.126 nm. These also matched with the FCC copper lattice distances of 0.18 nm and 0.128 nm for the (200) and (220) planes, respectively. It is confirmed from the SAED pattern that pure polycrystalline copper particles are synthesized in the DI water. It has been observed that due to the agglomeration of copper nanoparticles, the particles synthesized in pure DI water are of sub-micron size rather than nano size. Even though ultrasonic vibration was provided to avoid agglomeration, still large sized particles were observed in the obtained colloidal suspension. This indicates that ultrasonication was not so effective in preventing agglomeration of particles during the experiment. Moreover, poor dispersion stability with sedimentation of copper particles in DI water was noticed by visual inspection after 4 hr of the collection of the obtained Cu-DI water colloidal suspension. To overcome this problem, as per the surface chemistry of copper particle and DI water, Poly-Vinyl Alcohol (PVA) and Poly-Ethylene Glycol (PEG) are used as polymeric stabilizers to study their effect on the interaction between the nucleated synthesized copper nanoparticles. These stabilizers exhibit exemplary characteristics such as non-ionic, water soluble, film forming qualities, and wetting effect.

5.3.1.2 Size, Shape, and Distribution in DI Water + PVA Solution

Poly-Vinyl Alcohol (PVA) was added to the DI water at concentration of 1 wt/vol %. Characterization was carried out on the Cu-DI water + PVA colloidal suspension sample collected from the experiment conducted by following a similar procedure and under the same input conditions as given in Table 5.1. The PVA used has a molecular weight of 9000. From the TEM micrograph as shown in Figure 5.1d it was analyzed that the copper nanoparticles synthesized in DI water + PVA solution are spherical in shape and well dispersed with a small broad size distribution (range: 2–10 nm and mean diameter: 7.04 nm) Figure 5.1e). The particle size measurement is found to follow log-normal distribution. The selected area electron diffraction pattern in the inset (Figure 5.1d) showed a bright ring pattern with a d spacing of 0.21 nm that matched perfectly with the FCC crystalline copper lattice distance of 0.208

nm for the (111) plane. Similarly, the lattice spacing for the next two rings matched with the established lattice spacing of copper for the (200) and (220) planes. However, the diffraction pattern shows amorphous diffuse scattering rings rather than clear diffraction rings. This is attributed to the amorphous nature of stabilizer present in the peripheral regions of the particles. The EDAX pattern of the copper nanoparticles in DI water + PVA solution is shown in Figure 5.1f. The EDAX analysis was carried out over a small area of the precipitate rather than a single particle. This is due to the size of the particles being very small and PVA stabilizer capping the particles, so analysis of a single particle became too difficult. The analysis also showed the characteristic peaks of elemental copper identifying the presence of copper nanoparticles synthesized in the solution. The carbon and oxygen peak originated from the stabilizer present in the DI water. Also, impurities such as iron (Fe) and other negligible impurities were noticed in EDAX that cannot be excluded.

The size of the particles synthesized using DI water + PVA solution was certainly reduced as observed in Figure 5.1d by counteracting the forces of attraction owing to high surface energy between the nanoparticles. The mechanism of counteracting the forces of attraction is when nanoparticles are dispersed in the base fluid, the dissolved stabilizer molecules arrange themselves at the interface between solid particles and liquid because of their amphiphilic nature. They are arranged in such a way that the hydrocarbon chain is in contact with the surface of the particle due to its characteristic chemical affinity while the polar head group is oriented towards the liquid phase. Therefore, the particle surface is surrounded by a thin layer of stabilizer chains. This polymeric shell forms a barrier that prevents close contact between particles and inhibits agglomeration.

However, by visual inspection sedimentation of nanoparticles was still inherently taking place after 36 hr of the collection of the Cu-DI water + PVA colloidal suspension. This effect is due to the high concentration of hydroxyl groups; PVA polymer does not adsorb extensively at particles interfaces. Therefore, in this study one more stabilizer—Poly-Ethylene Glycol (PEG)—was used to study the aggregation and sedimentation of copper nanoparticles. The PEG used has a molecular weight of 7000.

5.3.1.3 Size, Shape, and Distribution in DI Water + PEG Solution

Poly-Ethylene Glycol (PEG) was added to the DI water at concentration of 1 wt/vol %. Characterization was carried out on the Cu-DI water + PEG colloidal suspension sample collected from the experiment conducted under the same input conditions as provided in Table 5.1. Analysis of the TEM image as illustrated in Figure 5.1g showed that the copper nanoparticles generated in the DI water and PEG mixture are spherical and very well dispersed with a narrow size distribution (range: 4–10 nm and mean diameter: 6.04 nm) (Figure 5.1h). The narrow size distribution indicates the approximate uniformity in the size of the particles obtained in the PEG colloidal solution. Further, the SAED and EDAX pattern of copper nanoparticles dispersed in the PEG sample was observed to be similar as in inset Figure 5.1d and Figure 5.1f. When added to the DI water, it was revealed that the PEG stabilizer prevented the agglomeration of the particles, and mean size of the generated copper nanoparticles was reduced compared to the mean size obtained in the DI water + PVA solution. Also the excellent dispersion stability of the Cu-DI water + PEG colloidal suspension without any particles settlement was observed for a longer duration.

Table 5.2 shows the size characterization of colloidal suspensions of copper nanoparticles. The summary of details of size

TABLE 5.2   Size Characterization of Colloidal Suspensions of Copper Nanoparticles

Colloidal suspension	Particles size range (nm)	Mean particle size (nm)
Copper-DI water	600–1100	880
Copper-DI water + PVA	2–10	7
Copper-DI water + PEG	4–10	6

range and mean size of the copper particles generated in pure DI water and DI water with stabilizers are also provided in Table 5.2.

5.3.2 Colloidal Aluminium Nanoparticles

The procedure to synthesize colloidal aluminium nanoparticles using the developed EDMM prototype is similar to those adopted for copper nanoparticles. Table 5.3 shows the experimental parameters used for the synthesis of aluminium nanoparticles. The size, shape, and distribution characterization of aluminium nanoparticles synthesized in the suspension media are also carried out in a similar manner as was carried out for colloidal copper nanoparticles, and the results are discussed in the following section.

5.3.2.1 Size, Shape, and Distribution in Pure DI Water

The TEM images of aluminium nanoparticles synthesized in pure DI water is shown in Figure 5.2a. The particles synthesized are near circular in shape with corrugated edge and some particles are of thin flake like structures. The image showed the presence of accumulated nano-aluminium clusters with defined singles. This shows that the high number density of the particles allows for their interaction in spite of their restricted mobility during thermal machining.

TABLE 5.3   Experimental Parameters Used for the Synthesis of Aluminium Nanoparticles

Parameters	Values
Open circuit voltage	20 V
Peak current	2 A
Frequency	4 kHz
Duty cycle	30%
Pulse duration	75 µs
Workpiece electrode (anode)	Aluminium plate of 460 µm thickness
Tool electrode (cathode)	Aluminium wire of 900 µm diameter
Dielectric fluid (base fluid)	De-Ionized (DI) water—200 ml
Environment	Room temperature

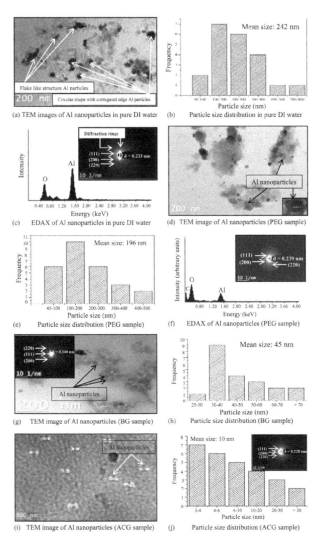

FIGURE 5.2 Size, shape, and distribution of colloidal aluminium nanoparticles: (a) TEM images of Al nanoparticles in pure DI water, (b) Particle size distribution in pure DI water, (c) EDAX of Al nanoparticles in pure DI water, (d) TEM image of Al nanoparticles (PEG sample), (e) Particle size distribution (PEG sample), (f) EDAX of Al nanoparticles (PEG sample), (g) TEM image of Al nanoparticles (BG sample), (h) Particle size distribution (BG sample), (i) TEM image of Al nanoparticles (ACG sample), (j) Particle size distribution (ACG sample).

The size of the particles was measured and implemented for size distribution analysis of particles. Figure 5.2b shows the size distribution of aluminium nanoparticles in pure DI water. The particles size lies in the range of 40 nm to 600 nm and the mean size of the particle was determined to be 242 nm. The particle size measurement is found to follow log-normal distribution. Figure 5.2c shows the EDAX pattern of aluminium particles in pure DI water. It is observed that the major content of the colloidal solution obtained is aluminium. As aluminium contains low atomic density crystallographic facets, one characteristic peak of aluminium particle was found to have appeared at a crystal facet. Also, the oxygen peak and additional peaks were observed indicating the presence of oxygen content and some inorganic impurities during machining. The SAED pattern in inset (Figure 5.2c) shows clear diffraction rings, with the first ring diameter from the centre yielding lattice (d) spacing of 0.233 nm, which perfectly matched with the crystalline aluminium lattice distance of 0.233 nm for the (111) plane. Similarly, the d spacing for the next two rings were found to be 0.201 nm and 0.142 nm. These also matched with aluminium lattice distances of 0.203 nm and 0.143 nm for the (200) and (220) planes, respectively. This information combined with the analysis from EDAX leads to the conclusion that the particles synthesized are crystalline aluminium with an amorphous aluminium oxide coating.

It was noticed that due to the agglomeration of nano-aluminium clusters followed by coalescence, the particles synthesized in pure DI water are of sub-micron size. In addition to that, poor stable dispersion with sedimentation of aluminium particles in DI water was observed visually after 12 hr of the collection of the obtained Al-DI water colloidal suspension. To overcome this problem, in the present study, as per the chemistry aspect, Poly-Ethylene Glycol (PEG), Bael Gum (BG), and ACacia Gum (ACG) are used as stabilizers to study their effect on the interaction between the nucleated aluminium nanoparticles synthesized.

5.3.2.2 Size, Shape, and Distribution in DI Water + PEG Solution

PEG was added to the DI water at concentration of 1 wt/vol % and the experiment was conducted under the same input conditions as provided in Table 5.3. Figure 5.2d–f shows the TEM image, histogram, and EDAX for Al-DI water + PEG colloidal suspension. Figure 5.2d shows that the particles are in the polyhedral shape and in the size range of 45 nm to 500 nm with mean size of 196 nm as shown in Figure 5.2e. Due to the polyhedral shape of the particles, diagonal distances were used as the descriptive size measurement for each particle. The synthesis of polyhedral aluminium particles may be attributed to the annealing of spherical particles formed during nucleation and particle growth. The EDAX pattern of polyhedral particle as shown in Figure 5.2f exhibits a clear peak of aluminium, rise in carbon, and oxygen intensity seems to have originated from the PEG stabilizer present in the DI water. The possible presence of other trace level impurities cannot be excluded. The SAED pattern as shown in inset (Figure 5.2f) showed a bright ring pattern with a d spacing of 0.239 nm that matched perfectly with the FCC crystalline aluminium lattice distance of 0.233 nm for the (111) plane. However, the diffraction pattern shows amorphous diffuse scattering rings rather than clear diffraction rings. This is attributed to the amorphous nature of stabilizer present in the peripheral regions of the particles.

It was observed that when PEG was added to the DI water, the mean size of the synthesized aluminium particles slightly reduced, but still large sized particles are produced due to clustering of particles. This indicates that the PEG polymer does not form a strong barrier to prevent close contact between aluminium particles. Also, by visual inspection sedimentation of nanoparticles was taking place after 24 hr of the collection of the Al-DI water + PEG colloidal suspension. To solve this problem, Bael Gum (BG)—a non-ionic polysaccharide—was employed as a stabilizer in the present study.

5.3.2.3 Size, Shape, and Distribution in DI water + BG Solution
The concentration of BG in DI water is taken as 1 wt/vol %. Characterization was done on the particles in the sample collected from the experiment conducted under the same input conditions as given in Table 5.3. Analysis of the TEM image as illustrated in Figure 5.2g showed that the aluminium nanoparticles synthesized in the DI water + BG solution are spherical and in the size range of 25 nm to 70 nm with mean size of 45 nm as shown in Figure 5.2h. The image clearly shows the surface adsorption of BG on the spherical aluminium nanoparticles. The SAED pattern in inset (Figure 5.2g) shows a series of electron diffraction spots and the crystal plane spacing (d spacing) of the inner core region was found to be 0.240 nm, which corresponds to the FCC crystalline aluminium d spacing of 0.233 nm for the (111) plane. It was proved from the SAED result that the particles synthesized in the BG sample are of pure crystalline aluminium. The BG when added to the DI water has certainly reduced the mean size of the synthesized aluminium particles, by reducing the interaction between the nucleated particles. But, by visual inspection sedimentation of nanoparticles was still taking place after 36 hr of the collection of the Al-DI water + BG colloidal suspension. Therefore, in this study one more stabilizer—ACacia Gum (non-ionic polysaccharide)—was used to study the agglomeration and sedimentation of nanoparticles.

5.3.2.4 Size, Shape, and Distribution in DI Water + ACG Solution
The concentration of ACG in DI water is taken as 1 wt/vol % and the experiment has been conducted under same input conditions as provided in Table 5.3. The TEM image in Figure 5.2i shows that the ACG colloidal suspension consists of very well dispersed spherical aluminium nanoparticles with a narrow size distribution in the size range of 3 nm to 30 nm, and mean size of 10 nm as shown in Figure 5.2j. The narrow size distribution indicates the approximate uniformity in the size of the particles obtained in the ACG colloidal solution as compared to the Al-DI water, PEG,

and BG colloidal suspensions. Further, the SAED pattern of aluminium nanoparticles as shown in the inset (Figure 5.2j) looks like an amorphous diffuse scattering ring, which was observed to be similar as in the inset (Figure 5.2f).

Thus, when the ACG polymeric stabilizer was added to the DI water, it was found that agglomeration of the particles was prevented, mean size of the aluminium nanoparticles was reduced compared to the mean size obtained in the DI water + BG solution and the sedimentation of nanoparticles was also reduced giving an excellent stable dispersion of the Al-DI water + ACG colloidal suspension for a longer time.

Table 5.4 shows the size characterization of colloidal suspensions of aluminium nanoparticles. The summary of details of size range and mean size of the aluminium particles synthesized in pure DI water and DI water with stabilizers are provided in the table.

5.4 CHEMICAL CHARACTERIZATION OF SYNTHESIZED NANOPARTICLES

5.4.1 Optical Absorption of Colloidal Copper Nanoparticles

The characteristic optical absorption spectra of colloidal copper nanoparticles are analyzed using the UltraViolet-visible (UV-vis) spectroscopy method. The analysis was carried out two weeks after the collection of the synthesized samples, to ascertain the chemical nature of the samples. It should be noted that the position and shape of the UV-vis absorption peaks are related

TABLE 5.4   Size Characterization of Colloidal Suspensions of Aluminium Nanoparticles

Nanofluid	Particles size range (nm)	Mean particle size (nm)
Aluminium-DI water	40–600	242
Aluminium-DI water + PEG	45–500	196
Aluminium-DI water + BG	25–70	45
Aluminium-DI water + ACG	3–30	10

to particle size, shape, inter-particle spacing, dielectric environment, and dielectric properties of the nanoparticles (Das et al., 2008; Chan et al., 2007; Bohren and Huffman, 1983). Figure 5.3a shows the UV-visible absorption spectrum of the copper-DI water colloidal suspension. The spectrum has a maximum absorbance peak of 1.3 at wavelength of 215 nm. It shows the existence of copper nanoparticles with a sign of the formation of a mono-oxide

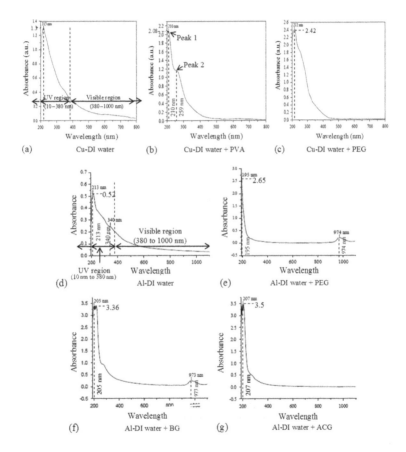

FIGURE 5.3   UV-visible absorption spectrum of colloidal suspensions of particles: (a) Cu-DI water, (b) Cu-DI water + PVA, (c) Cu-DI water + PEG, (d) Al-DI water, (e) Al-DI water + PEG, (f) Al-DI water + BG, (g) Al-DI water + ACG.

layer on the surface of the particles. This peak corresponds to the inter band transition from deep level electrons of valence band of cuprous oxide (Swarnkar et al., 2009). Tilaki et al. (2007) carried out the absorption spectra analysis of colloidal copper nanoparticles in DI water collected after two weeks of preparation and also reported the formation of an oxide layer over the surface of the nanoparticles.

The formation of an oxide layer over the surface of the copper nanoparticles inhibits the formation of the Surface Plasmon Resonance (SPR) peak in the visible region of the spectrum. The SPR is the collective oscillation (excitation) of the free conduction electrons caused by the interaction of light with the free electrons of the metallic particles.

Figure 5.3b shows the UV-visible absorption spectrum of the copper-DI water + PVA colloidal suspension. The spectrum shows two absorption peaks for the Cu-DI water + PVA colloidal suspension. The maximum absorption peak (peak 1) of 2.08 appeared at the wavelength of 210 nm. It also shows the existence of oxidized copper nanoparticles in the suspension. This may be attributed to weak adsorption of PVA molecules on the particles' surface because of high concentration of hydroxyl groups in PVA that leads to the copper nanoparticles reacting with dissolved oxygen in water and becoming oxidized. A quite similar observation has also been reported by Khanna et al. (2007). A weak intensity of absorption peak (peak 2) towards the higher wavelength was observed, and it was centred at 259 nm, which is related to the band edge electronic transition of oxidized copper nanoparticles dispersed in the DI water + PVA solution.

Figure 5.3c shows the UV-visible absorption spectrum of the copper-DI water + PEG colloidal suspension. The spectrum has a maximum absorbance peak of 2.42 at wavelength of 212 nm. It also shows the existence of a mono-oxide layer on the copper nanoparticles. This may be due to the presence of high concentration of ether groups in PEG and hence the copper nanoparticles react with oxygen atoms of the ether groups and become oxidized.

Due to the oxidation of copper nanoparticles in DI water + PVA and PEG, the SPR peak is also absent in the visible region of the spectrum. The intensity of maximum absorption peak in the Cu-DI water + PEG colloidal suspension was found to be higher than Cu-DI water and Cu-DI water + PVA suspensions in the UV range of the spectrum. This is attributed to the reduction in the size of the copper nanoparticles generated in the DI water + PEG mixture. It is also clear from Figure 5.3c that the absorption of light decreases at longer wavelengths (i.e. over the range of visible spectrum) as compared to the absorption spectrum observed in Figures 5.3b and c. This is attributed to the smaller size of the copper particles generated in the PEG sample which is a characteristic feature of metal particles of such sizes.

Moreover, in the UV-vis absorption spectrum of the Cu-DI water + PEG colloidal suspension, the absorption band at around 212 nm wavelength was observed to be narrow. A narrow absorption band clearly indicates a very narrow size distribution of copper nanoparticles within the PEG solution, which is in agreement with the TEM image and size distribution histogram as shown in Figures 5.1g and h.

5.4.2 Optical Absorption of Colloidal Aluminium Nanoparticles

The characteristic optical absorption spectra of colloidal aluminium nanoparticles were analyzed by the UV-vis spectroscopy method two weeks after of the collection of the generated samples. The UV-vis absorption spectra of colloidal aluminium nanoparticles at room temperature were recorded in the wavelength range of 190 nm to 1100 nm.

The maximum absorbance peak of 0.52 at wavelength of 213 nm as shown in Figure 5.3d corresponds to the localized surface plasmon resonance of Al nanoparticles in the middle UV range of spectrum (Chan et al., 2008). The shape of the absorption band demonstrates uneven distribution below the peak maximum. A weak absorption signal was observed towards the higher wavelength at 340 nm, which can be attributed to the oxidation of aluminium nanoparticles dispersed in DI water. In the Al-DI water + PEG colloidal suspension as shown in Figure 5.3e, the spectrum shows maximum absorption peak of 2.65 at the wavelength of 195 nm which is related to the Surface Plasmon Resonance (SPR) of Al nanoparticles in the middle UV range of spectrum. A weak peak located at 974 nm was due to the presence of thin oxide shell on the surface of generated aluminium particles. The maximum absorption peak for the Al-DI water + BG colloidal suspension of 3.36 as shown in Figure 5.3f appears to be at about 205 nm and for the Al-DI water + ACG colloidal suspension of 3.5 as shown in Figure 5.3g appears to be at about 207 nm. These peaks correspond to the plasmon peaks for colloidal aluminium. However, in Figure 5.3f, the existence of some weak signals at 270 nm and 973 nm are most likely due to the oxidation of colloidal aluminium nanoparticles in the BG sample. This may be an indication that in this case even though separated aluminium nanoparticles were almost produced, the adsorption of BG polymer on the surface of nanoparticles was not strong enough to avoid the oxidation. It was observed in Figures 5.3d–g that there is a shift in the wavelength of UV light for different colloidal suspensions of aluminium samples where maximum absorption takes place. This is due to the change in the size distribution, shape, and mean size of the colloidal aluminium particles.

The intensity of maximum absorption peak in the Al-DI water + ACG colloidal suspension was found to be higher than other colloidal suspension samples in the UV range of the spectrum. Apparently, this can attributed to the reduction in the size of the aluminium nanoparticles generated in the ACG sample. As the mean size of the particles generated in ACG solution was decreased, the absorption spectrum of visible light as shown in Figure 5.3g slightly decreases at longer wavelengths as compared to the spectrums appearing in Figures 5.3d and e; a characteristic feature of metal particles of such reduced sizes. Furthermore, in the UV-vis absorption spectrum of the Al-DI water + ACG colloidal suspension, a narrow plasmonic band was observed at around 207 nm wavelength as compared to other samples. This clearly indicates the generation of small size nanoparticles with a very narrow size distribution within the ACG solution, which is in agreement with the TEM image and size distribution histogram as shown in Figures 5.2i and j, respectively.

5.5 APPLICATION CHARACTERIZATION OF SYNTHESIZED NANOPARTICLES

5.5.1 Concentration Measurement of Colloidal Nanoparticles

The concentration measurement of colloidal copper suspension of nanoparticles has been carried out using the ultrasonic velocity technique, which is discussed in Section 3.4.1, Chapter 3. The average acoustic time of flight for each suspension medium and colloidal suspension is determined using Equation 3.10. The maximum difference of the measured time of flight between successive positions of the probe in each sample is found to be less than 1%. Table 5.5 shows the average time of flight for different samples measured at 25°C.

The average acoustic time of flight (T_s) was used to calculate the ultrasonic velocity (v_s) in different samples according to Equation 3.11. Table 5.6 shows the ultrasonic velocity for different samples. It was

TABLE 5.5 Average Time of Flight for Different Samples Measured at 25°C

Test samples	Time of flight, T_s (μs)
Suspension medium	
DI water	6.711
DI water + PVA	6.675
DI water + PEG	6.662
Colloidal suspensions	
Cu-DI water	6.505
Cu-DI water + PVA	6.477
Cu-DI water + PEG	6.483

TABLE 5.6   Ultrasonic Velocity
for Different Samples

Test samples	Ultrasonic velocity (m/s)
DI water	1490
DI water + PVA	1498
DI water + PEG	1501
Cu-DI water	1537
Cu-DI water + PVA	1544
Cu-DI water + PEG	1542

found that the ultrasonic velocity in colloidal suspensions of copper nanoparticles increases as compared to the suspension media. The obvious reason is the presence of nanoparticles in the fluid that change its density and compressibility. The ultrasonic velocity (v_s) of colloidal suspension is determined using the Equations 3.5–3.7. The volume fraction (ϕ_1) of the nanoparticles is determined by substituting Equation (3.8) in Equation (3.5) and thus the mass fraction (w) of the colloidal suspension is determined using Equation (3.9).

In this study, the density of water was taken as 997.0 kg/m³ at 25°C (Marsh, 1987). Moreover, the density and compressibility of bulk copper are taken as the same for the copper nanoparticle. Table 5.7 shows the properties of suspension media and copper nanoparticle at 25°C. The properties of the suspension media and copper nanoparticle are used for the concentration calculation of colloidal copper nanoparticles.

TABLE 5.7   Properties of Suspension Media and Copper Nanoparticle at 25°C

Properties	Suspension media			Cu-nanoparticle
	DI water	DI water + PVA	DI water + PEG	
Density, ρ (kg/m³)	997.00	997.40	998.60	8940.00
Bulk modulus, β (Pa)	2.21×10^9	2.24×10^9	2.25×10^9	140×10^9
Compressibility, k (Pa⁻¹)	4.52×10^{-10}	4.46×10^{-10}	4.44×10^{-10}	7.14×10^{-12}

Table 5.8 shows the concentration characterization of colloidal copper nanoparticles. It can be observed that the concentration of copper nanoparticles in the DI water, DI water + PVA solution and DI water + PEG solution was found to differ slightly from each other. This may be attributed to the lack of identification of machining time while machining the workpiece in suspension media, however similar working conditions were maintained for all the experiments. The concentration measurement of colloidal aluminium nanoparticles generated in pure DI water and DI water with stabilizers was carried out in a similar manner as discussed for colloidal copper nanoparticles and the concentration results are provided in Table 5.9.

5.5.2 Thermal Conductivity Measurement

5.5.2.1 Colloidal Copper Nanoparticles

The properties of base fluid and the size, morphology, and concentration of particles and in addition the presence of stabilizers greatly affect the properties of colloidal suspensions

TABLE 5.8   Concentration Characterization of Colloidal Copper Nanoparticles

Colloidal suspensions	Concentration (mass fraction, %)
Cu-DI water	0.146
Cu-DI water + PVA	0.135
Cu-DI water + PEG	0.128

TABLE 5.9   Concentration Characterization of Colloidal Aluminium Nanoparticles

Nanofluid	Concentration (mass fraction, %)
Al-DI water	0.32
Al-DI water + PEG	0.09
Al-DI water + BG	0.1
Al-DI water + ACG	0.16

(Das et al., 2003). In the present study, similar working conditions were maintained for all the machining experiments so as to keep the nanoparticles' concentration in different samples to be almost constant. Thermal conductivity analysis of the suspension media and colloidal copper nanoparticles has been carried out using the apparatus which was discussed in Section 3.4.2, Chapter 3. In this study, at least three thermal conductivity readings were taken for each sample to calculate the mean value of the measured data. The maximum deviation of the measured data was observed to be less than 10%.

Table 5.10 shows the thermal conductivity of the suspension media and colloidal suspensions of copper at 20°C. The percentage variation of measured thermal conductivity of colloidal copper nanoparticles samples with DI water are presented in Table 5.10. The thermal conductivity enhancement of colloidal suspensions of copper was calculated from measured thermal conductivity data as follows:

$$\% \text{ Increase} = \frac{K_{nf} - K_{bf}}{K_{bf}} \times 100 \quad (5.1)$$

where K_{nf} and K_{bf} are the thermal conductivities of colloidal suspensions and base fluid (DI water), respectively.

TABLE 5.10   Thermal Conductivity of Suspension Media and Colloidal Suspensions of Copper at 20°C

Test sample	Thermal conductivity (W m^{-1} K^{-1})	% Decreased (compared to DI water)
DI water	0.572	–
DI water + PVA	0.550	4
DI water + PEG	0.560	2
		% Increased (compared to DI water)
Cu-DI water	0.606	6
Cu-DI water + PVA	0.649	14
Cu-DI water + PEG	0.656	15

From Table 5.10 it was observed that the increase in thermal conductivity of copper-DI water colloidal suspension, copper-DI water + PVA colloidal suspension and copper-DI water + PEG colloidal suspension was almost 6%, 14%, and 15% with respect to pure DI water. The increase in thermal conductivity of colloidal suspensions of copper with respect to DI water is due to the suspension of nanoparticles in the base fluid and greater surface area due to decrease in particle size (Vadasz, 2006). On the other hand, additional factors that are also responsible for thermal conductivity enhancement includes Brownian motion of nanoparticles, liquid layering at the liquid/particle interface, strong movement of energy carriers within nanoparticles and nanoparticles structure formation via fractal, clumping and networking (Kathiravan et al., 2010; Jang and Choi, 2004). In reality, stabilizers considerably affect the heat transfer of fluids as a result of solely altering their surface tension and wettability (Kathiravan et al., 2010; Vassallo et al., 2004). From Table 5.10, it is also observed that on the addition of stabilizers thermal conductivity of DI water decreases and this is because of the decrease in surface tension and increase in wettability of the base fluid being affected by the stabilizer. But after the machining process the thermal conductivity of DI water and DI water with stabilizers is found to have increased and this is attributed to the copper nanoparticles suspended in the fluid. Thermodynamic penalty has to be paid with the use of stabilizers, as there was a reduction in the thermal conductivity on their addition to the base fluid.

5.5.2.2 Colloidal Aluminium Nanoparticles

Table 5.11 shows the thermal conductivity of suspension media and colloidal suspensions of aluminium at 25°C. The percentage variation of measured thermal conductivity of colloidal aluminium nanoparticles samples with DI water are presented in the table. From Table 5.11 it was observed that the enhancement in thermal conductivity of Al-DI water, Al-DI water with PEG, BG, and ACG colloidal suspensions was almost 5%, 8%, 7%, and 9%

TABLE 5.11   Thermal Conductivity of Suspension Media and
Colloidal Suspensions of Aluminium at 25°C

Test sample	Thermal conductivity ($W\ m^{-1}\ K^{-1}$)	% Decreased (compared to DI water)
DI water	0.609	
DI water + PEG	0.599	2
DI water + BG	0.604	1
DI water + ACG	0.606	0.5
		% Increased (compared to DI water)
Al-DI water	0.641	5
Al-DI water + PEG	0.654	8
Al-DI water + BG	0.650	7
Al-DI water + ACG	0.663	9

with respect to pure DI water. The enhancement in thermal conductivity of colloidal suspensions of aluminium with respect to
DI water is due to the suspension of nanoparticles in the base fluid
and larger surface area to volume ratio due to decrease in particle
size. In contrast, other mechanisms that are also responsible for
thermal conductivity enhancement include Brownian motion of
the nanoparticles, liquid layering at the liquid/particle interface,
nature of the heat transport in the nanoparticles, and nanoparticles clustering (Keblinski et al., 2002, Keblinski et al., 2005). From
Table 5.11, it is also observed that on the addition of stabilizers
thermal conductivity of DI water decreases and this is attributed
to the decrease in surface tension and increase in wettability of
the base fluid being affected by the stabilizer. However, after the
machining process the thermal conductivity of DI water and DI
water with stabilizers is found to have increased because of the
suspension of aluminium nanoparticles in the fluid.

5.5.3 Viscosity Measurement

5.5.3.1 Colloidal Copper Nanoparticles

The viscosity analysis of the suspension media and colloidal
copper nanoparticles has been carried out using the Brookfield

viscometer which was discussed in Section 3.4.3, Chapter 3. The viscometer was calibrated using silicon oil as standard fluid. Care should be taken to attach the spindle to the viscometer by raising the sleeve. The spindle is rotated in the test samples contained in an Ultra Low Adapter at a selected speed of 120 rpm. The viscosity readings of each sample are recorded for a fixed period of 3 minutes. The viscosity of the test samples was measured under the consideration of laminar flow conditions. Table 5.12 shows the viscosity of suspension media and colloidal suspensions of copper at 25°C. The percentage variation of measured viscosity of colloidal copper nanoparticles samples with DI water are presented in the table.

From Table 5.12, it was observed that the viscosity of DI water increases with the addition of stabilizers and this is because of the long polymer chain molecules that fill the network with water forming a gel-like structure. But as the molecular weight of PVA (molecular weight of 9000) is higher than PEG (molecular weight of 7000), the viscosity of the DI water + PEG solution was found to be less than the DI water + PVA solution. The increase in viscosity of copper-DI water, copper-DI water + PVA and copper-DI water + PEG colloidal suspensions was almost observed to be 25%, 58%, and 63% with respect to pure DI water. This is due to the presence of nanoparticles which has a large impact on the potential enhancement. Although the viscosity of the PEG solution was

TABLE 5.12   Viscosity of Suspension Media and Colloidal Suspensions of Copper at 25°C

Test samples	Viscosity (mPa-s)	% Increased (compared to DI water)
DI water	1.06	–
DI water + PVA	1.58	49
DI water + PEG	1.50	42
Cu-DI water	1.32	25
Cu-DI water + PVA	1.67	58
Cu-DI water + PEG	1.73	63

less, the viscosity enhancement of the PEG colloidal suspension was seen to be more than the PVA colloidal suspension. This is attributed to the reduction in the size of the particles synthesized in the PEG solution.

It may be noted that the viscosity of base fluids seems to be an important factor affecting the dispersion of nanoparticles and the stability of suspensions (Xuan and Li, 2000). The increase in viscosity of the base fluid reduces Brownian diffusion (i.e. reduces the diffusion of dispersed phase material), which in turn reduces the rate of coalescence followed by the reduction in collision rate, and thus stabilization can be possibly achieved (Challis et al., 2005). So, in this study, the dispersion stability of colloidal suspensions was evaluated using Stokes law of viscosity by the measured viscosity values of base fluid and base fluid with stabilizers.

According to Stokes law of viscosity, the sedimentation velocity (V) in m/s follows:

$$V = \frac{gD^2(\rho_p - \rho_f)}{18\mu_d} \tag{5.2}$$

where μ_d is dynamic viscosity of the suspension medium (mPa-s), g acceleration due to gravity (m/s^2), D size of the particle (nm), ρ_p density of the particle (kg/m^3), and ρ_f density of fluid (kg/m^3). Equation 5.2 is used to determine the relation between the sedimentation velocities of colloidal suspensions of copper samples, thus determining the dispersion stability.

The ratio $V_{PVA}/V_{DI\ water}$, $V_{PEG}/V_{DI\ water}$, and V_{PEG}/V_{PVA} could be determined using the viscosity values of suspension media (Table 5.12), mean size values of copper nanoparticles (Table 5.2), density of copper nanoparticle, base fluid, and base fluid with stabilizers PVA and PEG which are found to be 8940 kg/m^3, 997 kg/m^3, and 997.4 kg/m^3 (DI water + PVA); 998.6 kg/m^3 (DI water + PEG) as given in Table 5.7. Hence the ratios are found to be: 0.0098, 0.0001, and 0.775.

Here $V_{DI\ water}$, V_{PVA}, and V_{PEG} are the sedimentation velocities of copper nanoparticles in copper-DI water, copper-DI water + PVA and copper-DI water + PEG colloidal suspensions, respectively. It was found that from ratios 0.0098 and 0.0001, there is a reduction of 98.8% and 99.9% in the sedimentation velocity of copper nanoparticles in copper-DI water with PVA and PEG colloidal suspension when compared to copper-DI water colloidal suspension. It was also found that from the ratio 0.775, there is a reduction of 22.5% in the sedimentation velocity of copper nanoparticles in copper-DI water with PEG colloidal suspension when compared to copper-DI water with PVA colloidal suspension. Thus, there is an excellent decrease in the sedimentation of copper nanoparticles when PEG is added to the dielectric rather than PVA. Visual inspection of the colloidal suspension samples also confirms that Cu-DI water with PEG colloidal suspension displayed excellent dispersion stability without any sedimentation of particles for a longer duration of time. Hence, it is observed that the dispersion stability of Cu-DI water with PEG colloidal suspension as compared to Cu-DI water with PVA colloidal suspension is quite increased by changing the polymer from PVA to PEG. The better stability of nanofluids will prevent rapid settling and reduce clogging in the walls of heat transfer devices.

5.5.3.2 Colloidal Aluminium Nanoparticles

The viscosity analysis of the suspension media and colloidal aluminium nanoparticles has been carried out in a similar manner as was carried out in colloidal copper nanoparticles. Table 5.13 shows the viscosity of the suspension media and colloidal suspensions of aluminium at 30°C. The percentage variation of measured viscosity of colloidal aluminium nanoparticles samples with DI water are presented in the table. From Table 5.13, it was observed that the viscosity of DI water increases with the addition of stabilizers and this is attributed to the long polymer chain molecules that fill the network with water forming a rigid gel-like structure.

The increase in viscosity of Al-DI water, Al-DI water with PEG, BG and ACG nanofluids was almost observed to be 40%, 58%, 73%, and 78% with respect to pure DI water. This is due to the

TABLE 5.13   Viscosity of Suspension Media and
Colloidal Suspensions of Aluminium at 30°C

Test samples	Viscosity (mPa-s)	% Increased (compared to DI water)
DI water	1.01	–
DI water + PEG	1.32	31
DI water + BG	1.62	60
DI water + ACG	1.65	63
Al-DI water	1.41	40
Al-DI water + PEG	1.59	58
Al-DI water + BG	1.75	73
Al-DI water + ACG	1.80	78

presence of nanoparticles which has a large impact on the potential enhancement. The viscosity of the ACG colloidal solution was found to be higher than the other colloidal suspension samples and can be attributed to the decrease in the size of the particles synthesized in the ACG solution.

To evaluate the dispersion stability of colloidal suspensions of aluminium, the relation between the sedimentation velocities of nanofluid samples was determined using Equation 5.2. The ratio $V_{PEG}/V_{DI\ water}$, $V_{BG}/V_{DI\ water}$, and $V_{ACG}/V_{DI\ water}$, and V_{ACG}/V_{PEG} and V_{ACG}/V_{BG} could be determined using the viscosity values of suspension media (Table 5.13), mean size values of aluminium nanoparticles (Table 5.4), density of aluminium nanoparticle, base fluid, and base fluid with stabilizers PEG, BG, and ACG which are taken as 2700 kg/m^3, 995.6 kg/m^3, 996.0 kg/m^3 (DI water + PEG); 998.4 kg/m^3 (DI water + BG); 997.2 kg/m^3 (DI water + ACG). Hence the ratios are found to be: 0.5019, 0.0215, 1.045 × 10^{-3}, and 2.0824 × 10^{-3} and 0.0484. Here $V_{DI\ water}$, V_{PEG}, V_{BG}, and V_{ACG} are the sedimentation velocities of aluminium nanoparticles in aluminium-DI water, aluminium-DI water with PEG, BG, and ACG colloidal suspensions, respectively.

It was found that from ratios 0.5019, 0.0215, and 1.045 × 10^{-3}, there is a reduction of 49.8%, 97.8%, and 99.89% in the sedimentation velocity of aluminium nanoparticles in aluminium-DI water with PEG, BG, and ACG colloidal suspensions when compared to aluminium-DI water colloidal suspension. It was also found that from the ratio 2.0824×10^{-3} and 0.0484, there is a reduction of 99.79% and 95.15% in the sedimentation velocity of aluminium nanoparticles in aluminium-DI water with ACG colloidal suspension when compared to aluminium-DI water with PEG and BG colloidal suspensions. Hence, the ACG colloidal solution of aluminium nanoparticles showed excellent dispersion stability with a decrease in the sedimentation of particles as compared to other colloidal suspension samples, and also confirmed by visual inspection of the samples.

5.6 HEAT TRANSFER APPLICATION OF COLLOIDAL SUSPENSIONS

It is interesting to note that the applications of nanoparticles in various domains have specifically different demands, and thus face very different challenges. In the current scenario, there is a strong need for high performance heat transfer fluid to face the thermal challenges arising in many industries. However, conventional fluids used as coolants in thermal management systems so far exhibit a poor heat transfer rate. Although millimeter or micro sized particles-in-liquid suspensions are frequently used in industry, they are not suitable for heat transfer applications because of the rapid settling of these particles, damaging the walls of the pipe, large increases in pressure drop, and clogging. Therefore, in this study an innovative concept is illustrated to explore the possibilities of colloidal nanoparticles to achieve high performance heat transfer in thermal management systems.

The schematic diagram and photographic view of a developed thermal management system is shown in Figures 5.4a and b. The system is used to study the variation in temperature when the synthesized colloidal nanoparticles pass from source to sink. It consists of a stainless steel cylindrical vessel (capacity—120 ml) containing colloidal suspensions, a heater of 500 W capacity, brass

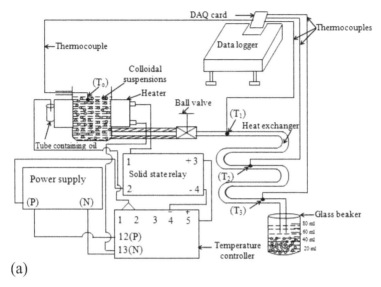

(a)

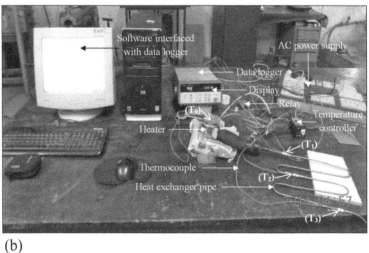

(b)

FIGURE 5.4   Thermal management system: (a) Schematic diagram, (b) Photographic view.

type ball valve, heat exchanger pipe of 3 mm diameter and 1500 mm length, solid state relay, temperature controller, T-type thermocouples to record the temperatures at various locations, and a glass beaker to collect the test samples after passing through the heat exchanger.

The temperature is recorded through the data logger interfaced with the computer system. The thermocouples are calibrated at ice point (0°C) and at boiling point of water (100°C). In the setup, the temperature variation is measured at different locations, which are denoted as T_1, T_2, and T_3 for every 10 seconds. The colloidal suspensions are heated to 40°C (T_0) and allowed to pass through the heat exchanger by opening the ball valve, and then the respective temperatures at different locations are recorded. Tables 5.14 and 5.15 show the temperature variation of DI water and colloidal suspensions of copper and aluminium particles at three locations.

It has been observed that the temperature of Cu-DI + PEG and Al-DI + ACG colloidal suspensions was found to have decreased as compared to DI water. A similar observation was also noticed by heating the colloidal suspensions to 45°C and allowing to pass through the heat exchanger. This indicates that colloidal suspensions of copper and aluminium have the ability to transfer more heat in the existing heat transfer system when compared to the base fluid of DI water which was employed as a traditional heat

TABLE 5.14   Temperature Variation of DI Water and Colloidal Suspensions of Copper Particles at Three Locations

Test fluid	Time stamp (min:sec)	T_0 (°C)	T_1 (°C)	T_2 (°C)	T_3 (°C)
DI water	23:26	40.1	39.2	37.9	36
	23:36	40.4	39.3	38.3	36.6
	23:46	40.6	39.1	38.7	37.7
	23:56	40.7	38.9	38.8	38.5
Cu-DI water + PEG	41:14	40.1	32.8	31.2	30.2
	41:24	40.5	37	34.6	32
	41:34	40.7	38.3	37	35.2
	41:44	40.8	38.5	37.6	36.7

transfer fluid. Further, from Tables 5.14 and 5.15 it was found that the colloidal suspension of copper has the ability to transfer more heat in the existing heat transfer system when compared to aluminium colloidal suspension. Thus colloidal suspensions of copper can be used as an effective heat transfer fluid in industries, particularly in automobiles and electronics by replacing the currently used conventional fluids.

5.7 SUMMARY AND CONCLUSIONS

The synthesis of copper and aluminium nanoparticles in the dielectric medium was carried out using the indigenously developed EDMM prototype system. The characterization of colloidal suspensions through various diagnostic studies has been studied. The TEM studies show that the size of the copper nanoparticles dispersed in pure DI water, DI water + PVA, and PEG solutions lies in the range of 600 nm to 1100 nm (mean size: 880 nm), 2 nm to 10 nm (mean size: 7.04 nm), and 4 nm to 10 nm (mean size: 6.04 nm), respectively. Similarly, the size range of aluminium nanoparticles generated in pure DI water, DI water + PEG, BG, and ACG solutions lies between 40 nm to 600 nm (mean size: 242 nm), 45 nm to 500 nm (mean size: 196 nm), 25 nm to 70 nm (mean size: 45 nm), and 3 nm to 30 nm (mean size: 10 nm), respectively. The EDAX and SAED results presented show the

TABLE 5.15   Temperature Variation of DI Water and Colloidal Suspensions of Aluminium Particles at Three Locations

Test fluid	Time stamp (min:sec)	T_0 (°C)	T_1 (°C)	T_2 (°C)	T_3 (°C)
DI water	23:26	40.1	39.2	37.9	36
	23:36	40.4	39.3	38.3	36.6
	23:46	40.6	39.1	38.7	37.7
	23:56	40.7	38.9	38.8	38.5
Al-DI water + ACG	20:21	40.1	37.2	34.8	32.3
	20:31	40.2	38.1	37.4	35.8
	20:41	40.2	38.7	38.3	37.9
	20:51	40.3	38.8	38.5	38.1

presence of crystalline copper and aluminium particles. The study on the size distribution of generated copper and aluminium particles indicates that the measurement follows log-normal distribution.

The concentration of copper nanoparticles dispersed in DI water, DI water + PVA, and DI water + PEG mixture was determined to be 0.146, 0.135, and 0.128 mass fraction (%), respectively. Similarly, the concentration of aluminium nanoparticles was determined to be 0.32, 0.09, 0.1, and 0.16 mass fraction (%), respectively. The experimental results have revealed that the thermal conductivity of Cu-DI water, with PVA and with PEG colloidal suspensions has enhanced by 6%, 14%, and 15% with respect to pure DI water (0.572 W m^{-1} K^{-1}). Similarly, the thermal conductivity of Al-DI water colloidal suspension, Al-DI water with PEG, BG, and ACG colloidal suspensions was measured to be 0.641 W m^{-1} K^{-1}, 0.654 W m-1 K-1, 0.650 W m^{-1}K$^-$1, and 0.663 W m^{-1} K^{-1}, respectively. It was found that the thermal conductivity of these colloidal suspensions of aluminium has almost increased by 5%, 8%, 7%, and 9% with respect to pure DI water (0.609 W m^{-1} K^{-1}). The viscosity of Cu-DI water, with PVA and with PEG colloidal suspensions was estimated to be 25%, 58%, and 63% higher than that of pure DI water (1.06 mPa-s). The viscosity of Al-DI water colloidal suspension, Al-DI water with PEG, BG, and ACG colloidal suspensions was enhanced by 40%, 58%, 73%, and 78% with respect to pure DI water (1.01 mPa-s). Using PEG and ACG, an excellent stable dispersion of colloidal suspensions of copper and aluminium was found without any particles settlement for a longer time. This would prevent the settling and clogging of the particles in the walls of heat transfer devices. Heat transfer application of colloidal suspensions was studied using a developed thermal management setup. The results show that the colloidal suspension of copper has the ability to transfer more heat in the existing heat transfer system when compared to colloidal suspension of aluminium.

5.8 SCOPE FOR FUTURE WORK

Many industries have a strong need for nanoparticles in the form of or without colloid, which offer several benefits and would be significant to various industries. Although nanoparticles offer promising opportunities, there are still a number of technical problems on the road to commercialization. To overcome these problems, research on nanoparticles synthesis with uniform size and distribution, and high yield is needed. Some future direction on these challenging areas is highlighted:

• Study on the required size, shape, and distribution of nanoparticles with process optimization and reduced diffusion limited growth.

• Modeling (either mathematical or theoretical or finite element, etc.) to ascertain size, shape, and distribution of nanoparticles with available process parameters such as voltage, current, frequency, and duty cycle.

• Search for other advanced mechanical non-traditional micromachining techniques for nanoparticles synthesis with uniform size and distribution.

• On-line measurement of size, shape, distribution, and concentration of nanoparticles.

REFERENCES

Bohren, C.F. and Huffman, D.R. *Absorption and Scattering of Light by Small Particles*, USA: Wiley, 1983.

Challis, R.E., Povey, M.J.W., Mather, M.L. and Holmes, A.K. (2005) Ultrasound techniques for characterizing colloidal dispersions. *Reports on Progress in Physics*, 68, 1541–1637.

Chan, G.H., Zhao, J., Hicks, E.M., Schatz, G.C. and Duyne, R.P.V. (2007) Plasmonic properties of copper nanoparticles fabricated by nanosphere lithography. *Nano Letters*, 7 (7), 1947–1952.

Chan, G.H, Zhao, J., Schatz, G.C. and Duyne, R.P.V. (2008) Localized plasmon resonant spectroscopy of triangular aluminium nanoparticles. *Journal of Physical Chemistry C*, 112, 13958–13963.

Das, S.K., Choi, S.U.S., Yu, W. and Pradeep, T. *Nanofluids: Science and Technology*, USA: Wiley Interscience, 2008.

Das, S.K., Putra, N. and Roetzel, W. (2003) Pool boiling characteristics of nanofluids. *International Journal of Heat and Mass Transfer*, 46, 851–862.

Jang, S.P. and Choi, S.U.S. (2004) Role of Brownian motion in the enhanced thermal conductivity of nanofluids. *Applied Physics Letters*, 84, 4316–4318.

Kathiravan, R., Kumar, R., Gupta, A. and Chandra, R. (2010) Preparation and pool boiling characteristics of copper nanofluids over a flat plate heater. *International Journal of Heat and Mass Transfer*, 53, 1673–1681.

Keblinski, P., Eastman, J. and Cahill, D. (2005) Nanofluid for thermal transport. *Materials Today*, 8, 36–44.

Keblinski, P., Phillpot, S.R., Choi, S.U.S. and Eastman, J.A. (2002) Mechanisms of heat flow in suspensions of nano-sized particles (nanofluids). *International Journal of Heat and Mass Transfer*, 45, 855–863.

Khanna, P.K., Gaikwad, S., Adhyapak, P.V., Singh, N. and Marimuthu, R. (2007) Synthesis and characterization of copper nanoparticles. *Materials Letters*, 61, 4711–4714.

Marsh, K.N. *Standard Density of Water (Recommended Reference Materials for the Realization of Physicochemical Properties)*, UK: Blackwell Scientific Publications, 1987.

Swarnkar, R.K., Singh, S.C. and Gopal, R. (2009) Synthesis of copper/copper oxide nanoparticles: optical and structural characterizations. *Proceedings of American Institute of Physics*, 205–210.

Tilaki, R.M., Irajizad, A. and Mahdavi, S.M. (2007) Size, composition and optical properties of copper nanoparticles prepared by laser ablation in liquids. *Applied Physics A: Materials Science and Processing*, 88, 415–419.

Vadasz, P. (2006) Heat conduction in nanofluid suspensions. *Journal of Heat Transfer*, 128, 465–477.

Vassallo, P., Kumar, R. and Amico, S.D. (2004) Pool boiling heat transfer experiments in silica-water nano fluids. *International Journal of Heat Mass and Transfer*, 47, 407–411.

Xuan, Y. and Li, Q. (2000) Heat transfer enhancement of nanofluids. *International Journal of Heat and Fluid Flow*, 21, 58–64.

Index